Cyber Security Kill Chain – Tactics and Strategies

Breaking down the cyberattack process and responding to threats

Gourav Nagar

Shreyas Kumar

‹packt›

Cyber Security Kill Chain – Tactics and Strategy

Portfolio Director: Vijin Boricha

Relationship Lead: Prachi Sawant

Program Manager and Growth Lead: Ankita Thakur

Project Manager: Gandhali Raut

Content Engineer: Sujata Tripathi

Technical Editor: Arjun Varma

Copy Editor: Safis Editing

Proofreader: Sujata Tripathi

Indexer: Hemangini Bari

Production Designer: Alishon Falcon

First published: May 2025

Production reference: 2311025

Published by Packt Publishing Ltd.

Grosvenor House

11 St Paul's Square

Birmingham

B3 1RB, UK

ISBN 978-1-83546-609-4

www.packtpub.com

To Texas A&M University, a beacon of knowledge and innovation, and to the enduring Aggie spirit that unites the authors. Aggie values of integrity, leadership, and excellence continue to inspire and guide us in every endeavor. This work is a tribute to the community that fosters growth, camaraderie, and the relentless pursuit of greatness.

Gig 'em, Aggies!

Foreword

In today's interconnected digital world, cyber-resilience is a strategic imperative vital to assure national security, economic stability, and personal privacy. As cyber threats grow in sophistication, organizations of all sizes must be equipped to protect, detect, and respond to attacks. Analysis of the Cyber Kill Chain has been an essential tool in the arsenal of threat responders for decades. However, all the stages of a modern cyberattack (reconnaissance, weaponization, delivery, exploitation, installation, command and control, and so on) have changed significantly, with the advent of new technologies such as artificial intelligence and an increase in attacker sophistication. A fresh look is needed to understand the modern cyberattack, and this book accomplishes exactly that.

More than a series of steps, the Cyber Kill Chain is a proactive strategy that helps defenders disrupt attacks early, shifting from a reactive stance to one that intercepts threats before they cause harm. As cyber threats evolve, defenses must follow suit. Nation-state actors, cybercriminals, and hacktivists are increasingly agile, employing complex tactics that outmatch traditional, static defenses. Organizations must adopt dynamic, adaptable security strategies to stay ahead. This book offers a comprehensive guide for understanding the stages of a cyberattack, enabling defenders to develop targeted defenses for each phase, making attacks more difficult to succeed.

Whether you are just entering the field of cybersecurity or are a seasoned professional, this book offers the knowledge and tools needed to defend against modern cyber adversaries. The insights shared here will help you better understand the anatomy of a cyberattack and, more importantly, how to defend against it. As the cyber threat landscape evolves, this book will be an indispensable resource for staying ahead of the curve.

I have known Shreyas for over a decade, and he has been an active contributor to the cyber community and made presentations at the RSA Conference. Together, Gourav and Shreyas bring a unique combination of practical industry experience and academic expertise, making this book an invaluable resource for both professionals and students.

Rohit Ghai

CEO, RSA Security

Contributors

About the authors

Gourav Nagar is the director of information security at BILL Holdings Inc, where he leads the information security engineering and security operations team. With over a decade of experience in cybersecurity, Gourav has built robust security programs across various domains, including security engineering, incident response, threat detection, infrastructure security, and digital forensics. His career includes key roles at industry leaders such as Uber, Apple, and EY. Gourav holds a master of science in management information systems from Texas A&M University and multiple industry certifications, including CISSP, CISM, CHFI, and GIAC **Certified Forensic Analyst (GCFA)**.

Shreyas Kumar is a professor of practice at the Department of Computer Science and Engineering, Texas A&M University, with over 22 years of experience across start-ups and Fortune 100 companies, including Adobe, Uber, and Oracle. Shreyas is an accomplished author and speaker in cybersecurity, with numerous academic publications and conference talks. He holds a master's degree in computer science from Texas A&M University and a bachelor's degree from the Indian Institute of Technology, Roorkee. He holds multiple security certifications, including CISSP, HCISPP, CISA, and PCIP.

About the reviewers

Tim Kiermaier has over 17 years of experience within IT and cybersecurity, with roles ranging from helpdesk and system administration to SOC analyst and security engineering. He currently serves as the head of information security. Armed with a master's degree in information technology from Charles Sturt University, Tim has experience in government, defense system integrators, and not-for-profit, and he occasionally works for the Australian Cyber Security Centre.

I would like to thank all those who have helped indulge and nurture my curiosity for all things technology and information security. Special thanks to my wife, Tash, for keeping me focused and encouraging me to keep pushing forward into my next challenge.

Yugal Pathak "CyberYuvi" is a distinguished DFIR professional who possesses extensive experience in corporate and law enforcement domains. He specializes in digital forensics operations, cybercrime investigation, and analysis, with expertise in multimedia, mobile phone, and dark web forensics. With a strong background in techno-managerial proficiency, Yugal has successfully managed programs, projects, and teams across geographies, and he has contributed to R&D operations in digital forensics, including the development of indigenous forensic hardware and software products (for the Make in India initiative). He is a seasoned trainer, mentor, author, and speaker, having collaborated with law enforcement agencies, corporations, and academia.

I would like to express my heartfelt gratitude to my family and friends for their unwavering support and understanding, allowing me to dedicate time and effort to the ever-evolving field of digital forensics. I am deeply thankful to my mentors and colleagues who have provided me with invaluable knowledge, resources, and connections. Your contributions and guidance have been invaluable to me, and I am honored to be a part of this community.

Rambo Anderson-You is a dedicated red team operator and cybersecurity researcher with a rich background in enhancing the security of Active Directory and macOS environments. Currently performing red team adversary emulation projects at BILL, he has a proven track record in identifying and mitigating potential security threats across various platforms. He has over a decade of security experience, working in government and offensive security consulting. Rambo has had a lot of fun hacking car dealerships, banks, cloud providers, and telecoms (probably with the appropriate permission).

Girish Nemade, a cybersecurity professional with 12 years in consulting, extends his expertise to global BFSI clients. Currently a Senior Lead Cyber Security Architect for a leading global bank, Girish excels in leading, directing, and executing large-scale cybersecurity engagements, particularly in banking and finance. He has orchestrated large-scale global red team, purple team operations, and cyber tabletop engagements for CxO's. He has designed secure architecture blueprints and conducted threat modelling.

He holds certifications such as eNDP, CRTO, CRTP, CARTP, OSCP, OSWP, CEH, RHCE, ISO27001. He is passionate about cybersecurity and enjoys exploring new technologies, solving intricate problems and finding ways to secure it.

I am sincerely grateful to Packt Publishing for offering me the opportunity to review this book and be a part of this enriching experience. The chance to contribute to the review has not only broadened my understanding but also allowed me to delve deeper into the subject matter. I extend my heartfelt thanks to my family for their unwavering support and encouragement throughout this journey.

Table of Contents

5

Exploitation   81

6

Installation   105

7

Command and Control   121

8

Actions on Objectives   137

9

Cyber Security Kill Chain and Emerging Technologies   161

10

Legal and Ethical Aspects of the Cyber Security Kill Chain   187

11

The Future   227

12

A Proactive Approach   253

13

Unlock Your Exclusive Benefits   267

Preface

In an era where digital infrastructures underpin nearly every aspect of modern life, the rise of cyberattacks presents a significant threat to individuals, corporations, and governments. The cyber kill chain has emerged as a valuable framework that breaks down the anatomy of these cyberattacks, allowing cybersecurity professionals to detect, disrupt, and ultimately prevent malicious activities. The concept, originally developed by Lockheed Martin, is now widely used to understand and defend against **advanced persistent threats (APTs)** and other sophisticated cyber intrusions.

This book, *Cyber Security Kill Chain – Tactics and Strategies*, is designed to empower you with an in-depth understanding of the kill chain framework and the methodologies employed at each stage of a cyberattack. From Reconnaissance to Action on Objective, this comprehensive guide will walk you through the different phases attackers use to infiltrate systems, wreak havoc, and steal sensitive data. It also provides insights into defensive strategies designed to detect and mitigate threats at each stage, empowering you to defend your digital infrastructure proactively.

The book is a timely response to the increasing complexity of cyber threats. With the rapid growth of technologies such as **artificial intelligence (AI)**, the **Internet of Things (IoT)**, and quantum computing, the battlefield of cyber warfare is evolving faster than ever. Attackers are becoming more sophisticated, utilizing AI to automate attacks and bypass traditional defenses, while defenders must now implement cutting-edge technologies to keep up. This book aims to equip you with the knowledge and skills to protect against current threats and anticipate future challenges.

Each chapter is structured to provide a clear and detailed understanding of the specific phase in the cyber kill chain, supported by real-world case studies, practical tools, and advanced techniques. For instance, the chapter on **Weaponization** delves into how attackers create malicious payloads based on vulnerabilities discovered during **Reconnaissance**, offering practical strategies to detect and mitigate these threats. Similarly, the chapter on **Command and Control (C2)** highlights how attackers maintain covert communication with compromised systems and offer practical defensive measures that can be implemented to disrupt these activities.

Additionally, this book covers how emerging technologies such as AI are being integrated into both attack and defense strategies, signaling a future where cybersecurity must continuously adapt. The chapter on AI and the cyber kill chain teaches you how AI can predict attack vectors, monitor for anomalies, and even respond to real-time incidents. The growing role of AI in cybersecurity reflects a paradigm shift from reactive to proactive defense, helping defenders stay one step ahead of attackers.

There's a dedicated chapter to explore ethical and legal considerations, as the intersection of cybersecurity with laws such as GDPR and CCPA is becoming increasingly critical. As businesses grapple with increasing regulatory requirements, understanding these legal frameworks is essential for aligning cybersecurity strategies with compliance and ensuring that defensive measures uphold ethical standards.

This book will help you develop a comprehensive understanding of the cyber kill chain and gain the tools, techniques, and best practices needed to apply it in the real world. Whether you are a cybersecurity professional defending a large enterprise or a curious learner aiming to understand the intricacies of cyber warfare, this book offers valuable insights to bolster your defenses and anticipate future challenges.

Who this book is for

This book is written for cybersecurity professionals, IT administrators, network engineers, and students who seek a comprehensive understanding of modern cyber threats and defensive strategies. It is also valuable for business leaders and decision-makers who must understand the framework to make informed decisions about cybersecurity investments and strategies. While a basic understanding of cybersecurity principles is helpful, this book provides explanations that make it accessible to individuals with varying technical expertise.

Whether you are a cybersecurity beginner or an experienced practitioner looking to refine your skills, this book will equip you with the knowledge and tools necessary to understand and apply the cyber kill chain in real-world scenarios. Familiarity with fundamental cybersecurity concepts such as malware, phishing, and network security is recommended but not required, as the book covers these areas thoroughly within the context of the kill chain framework.

What this book covers

Chapter 1, *Understanding the Cyber Kill Chain*, introduces the cyber kill chain framework and its relevance in today's cybersecurity landscape. It discusses the various types of attackers, their motives, and the increasing importance of cybersecurity in defending against them.

Chapter 2, *Reconnaissance - The Initial Breach Plan*, explores the reconnaissance phase, where attackers gather information on their targets. Learn passive and active reconnaissance techniques and how defenders can proactively detect and prevent early-stage attacks.

Chapter 3, *Weaponization*, delves into the weaponization phase, where attackers create malicious payloads. Understand how vulnerabilities are exploited to craft malware, using case studies of famous attacks.

Chapter 4, *Delivery*, focuses on the delivery methods used by attackers, including phishing emails and drive-by downloads. It provides real-world case studies and strategies to block delivery attempts.

Chapter 5, Exploitation, teaches you how attackers exploit vulnerabilities to execute malicious code. The chapter covers various exploitation techniques and provides mitigation strategies to protect against them.

Chapter 6, Installation, explains how attackers establish control over compromised systems by installing malware. Understand defensive strategies for detecting and preventing unauthorized installations.

Chapter 7, Command and Control (C2), explores the techniques attackers use to maintain covert communication with compromised systems and how defenders can detect and disrupt these operations.

Chapter 8, Actions on Objectives, covers the final phase, during which attackers achieve their primary objectives, such as data exfiltration or system sabotage, and how defenders can respond to such activities.

Chapter 9, Cyber Kill Chain and Emerging Technologies, examines the integration of AI into the cyber kill chain, providing insights into how AI enhances detection, response, and prevention at every stage of the attack life cycle.

Chapter 10, Legal and Ethical Aspects of the Cyber Kill Chain, explores the legal and ethical considerations in implementing the cyber kill chain, emphasizing the importance of regulatory compliance and ethical cybersecurity practices.

Chapter 11, The Future, looks ahead to the future of cybersecurity, discussing emerging threats and innovations such as quantum computing and post-quantum cryptography.

Chapter 12, A Proactive Approach, advocates for shifting from a reactive to a proactive cybersecurity stance. It covers threat anticipation, incident preparedness, and fostering a cybersecurity-aware culture within organizations.

To get the most out of this book

To fully benefit from this book, it's helpful to have a basic understanding of cybersecurity concepts, such as malware, phishing, and network vulnerabilities. Familiarity with topics such as network traffic monitoring and intrusion detection systems will also enrich your learning experience. If you're new to the cyber kill chain, this book provides clear explanations to bring you up to speed while offering deeper insights for those with experience. Whether a student or a seasoned professional, this book will equip you with actionable knowledge and real-world case studies to strengthen your cybersecurity defenses.

Conventions used

The following is a text convention used throughout this book.

Bold: Indicates a new term, an important word, or words that you see on the screen. For instance, words in menus or dialog boxes appear in the text like this. For example: "Cybersecurity extends its reach to mobile devices, the **Internet of Things (IoT)**, cloud services, and emerging technologies such as artificial intelligence and quantum computing."

> Tips or important notes
> Appear like this.

Get in touch

Feedback from our readers is always welcome.

General feedback: If you have questions about any aspect of this book, email us at `customercare@packt.com` and mention the book title in the subject of your message.

Errata: Although we have taken every care to ensure the accuracy of our content, mistakes do happen. If you have found a mistake in this book, we would be grateful if you would report this to us. Please visit `www.packtpub.com/support/errata` and fill in the form. We ensure that all valid errata are promptly updated in the GitHub repository at `https://github.com/PacktPublishing/Cyber-Security-Kill-Chain---Tactics-and-Strategies`.

Piracy: If you come across any illegal copies of our works in any form on the internet, we would be grateful if you would provide us with the location address or website name. Please contact us at `copyright@packt.com` with a link to the material.

If you are interested in becoming an author: If there is a topic that you have expertise in and you are interested in either writing or contributing to a book, please visit `authors.packtpub.com`.

Share your thoughts

Once you've read *Cyber Security Kill Chain - Tactics and Strategies*, we'd love to hear your thoughts! Scan the QR code below to go straight to the Amazon review page for this book and share your feedback.

`https://packt.link/r/1835466095`

Your review is important to us and the tech community and will help us make sure we're delivering excellent quality content.

‹packt› _secpro

Stay relevant in a rapidly changing cybersecurity world– join 65,000+ SecPro subscribers

_secpro is the trusted weekly newsletter for cybersecurity professionals who want to stay informed about real-world threats, cutting-edge research, and actionable defensive strategies.

Each issue delivers high-signal, expert insights on topics like:

- Threat intelligence and emerging attack vectors
- Red and blue team tactics
- Zero Trust, MITRE ATT&CK, and adversary simulations
- Security automation, incident response, and more!

Whether you're a penetration tester, SOC analyst, security engineer, or CISO, **_secpro** keeps you ahead of the latest developments — no fluff, just real answers that matter.

Scan the QR code to subscribe for free and get expert cybersecurity insights straight to your inbox:

https://secpro.substack.com

Free Benefits with Your Book

This book comes with free benefits to support your learning. Activate them now for instant access (see the "*How to Unlock*" section for instructions).

Here's a quick overview of what you can instantly unlock with your purchase:

PDF and ePub Copies

Next-Gen Web-Based Reader

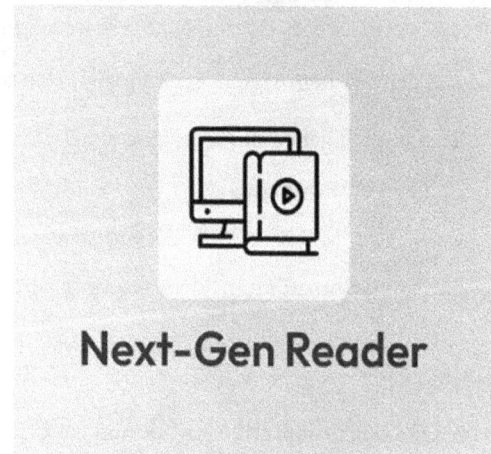

Access a DRM-free PDF copy of this book to read anywhere, on any device.

Use a DRM-free ePub version with your favorite e-reader.

Multi-device progress sync: Pick up where you left off, on any device.

Highlighting and notetaking: Capture ideas and turn reading into lasting knowledge.

Bookmarking: Save and revisit key sections whenever you need them.

Dark mode: Reduce eye strain by switching to dark or sepia themes.

How to Unlock

UNLOCK NOW

Scan the QR code (or go to `packtpub.com/unlock`). Search for this book by name, confirm the edition, and then follow the steps on the page.

Note: Keep your invoice handly. Purchase made directly from Packt don't require one.

Understanding the Cyber Security Kill Chain

In today's hyper-connected world, where global communication, commerce, and innovation are inseparable from digital infrastructure, cybersecurity is the critical guardian of our digital lives. Its role extends far beyond protecting personal data—cybersecurity is now fundamental to safeguarding industries, economies, and national security. As cyber threats evolve, cybersecurity's responsibilities have expanded dramatically, encompassing everything from securing sensitive information to defending the critical infrastructure that sustains entire nations.

At the forefront of this defense is cybersecurity, acting as a shield against a range of threat actor—cybercriminals, hacktivists, and state-sponsored entities—all intent on exploiting vulnerabilities for their gain. As the complexity and frequency of cyberattacks increase, the consequences of inadequate security measures grow more severe. The stakes of cybersecurity failures are higher than ever, from disruptions to essential services and catastrophic financial losses to heightened threats to national security. In today's digital age, robust cybersecurity is not just a priority; it is imperative for ensuring the stability and safety of our interconnected global systems. Every individual plays a crucial role in this collective defense.

The digital revolution has brought an era of unparalleled convenience and connectivity. We can now connect instantly with individuals across the globe, access vast amounts of information with a single click, and conduct business more efficiently than ever. However, this same interconnectedness has also introduced many risks and vulnerabilities.

Our growing dependence on interconnected devices, networks, and software has created a vast attack surface for cybercriminals to exploit. While the benefits of this connectivity are undeniable, it demands constant vigilance and proactive defense against the ever-evolving threats that target our digital world. In this ongoing battle, your vigilance is not just necessary—it's indispensable.

As cybersecurity professionals contend with various threat actors, from nation-states to hacktivists, understanding how to track and mitigate these threats is paramount. One guiding principle in this effort is **Locard's Exchange principle** from forensic science, which posits that any interaction leaves behind traces. In the cyber world, attackers inevitably leave behind digital footprints—whether in network logs, metadata, or files. This inevitability reassures us that we can predict and identify their actions. This forensic approach is critical when investigating breaches, tracing threat actors' movements, and implementing defensive strategies based on the evidence left behind.

Locard's Exchange Principle

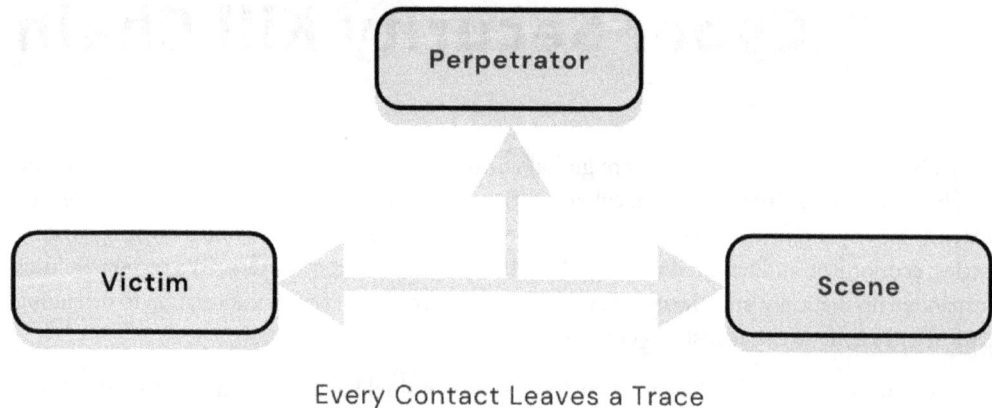

Perpetrator

Victim

Scene

Every Contact Leaves a Trace

Figure 1.1 – Locard's Exchange Principle

Data is the lifeblood of the modern world, with everything from personal information and intellectual property to critical infrastructure being stored and transmitted electronically. This wealth of data, while immensely valuable, has become a prime target for cybercriminals motivated by financial gain, political leverage, or the challenge of breaching security measures. As cyber threats intensify, cybersecurity emerges as a critical defense, employing various strategies, practices, and technologies to safeguard our digital assets and systems from relentless attacks.

Cybersecurity is proactive and adaptive at its core, continuously evolving to counter new threats and vulnerabilities. Security experts leverage their knowledge, advanced tools, and deep understanding of adversary tactics to stay ahead of malicious actors. The vast scope of cybersecurity encompasses various domains, including network security, which protects the flow of data; application security, which shields software from vulnerabilities; and information security, which ensures data confidentiality, integrity, and availability. Together, these disciplines form an essential defense in the digital age.

Cybersecurity extends its reach to mobile devices, the **Internet of Things (IoT)**, cloud services, and emerging technologies such as artificial intelligence and quantum computing. Each new technological advancement presents opportunities and challenges, demanding innovative security solutions.

Hacktivism, driven by political or ideological motives, often leverages the vulnerabilities found in IoT devices and human error to disrupt systems. A significant example of this is the 2016 Mirai botnet attack, where thousands of unsecured IoT devices were hijacked to launch a massive **Distributed Denial of Service (DDoS)** attack, crippling websites and online services globally. This demonstrated how the interconnected nature of IoT devices can be weaponized.

In parallel, the human element remains a persistent risk in cybersecurity. Hacktivist groups such as *Anonymous* have frequently exploited human vulnerabilities, such as weak passwords or susceptibility to phishing. A famous case involved the 2011 hack of HBGary Federal, where social engineering and poor password hygiene exposed sensitive emails and client data. IoT vulnerabilities and human error highlight the expanding and evolving nature of the threat landscape. While technology is at the forefront of cybersecurity, the human element is equally crucial. Cybersecurity awareness and education play a pivotal role in fortifying our defenses. Individuals, from the average internet user to top-level executives, must be cognizant of the risks they face in the digital realm and take proactive steps to protect themselves and their organizations.

In this chapter, we will cover the following key topics:

- The evolving landscape of cyber threats
- The significance of cybersecurity
- Types of threat actors and motives
- Introduction to the Cyber Kill Chain concept
- The state of cybersecurity in 2024
- The cost of a data breach

Free Benefits with Your Book

Your purchase includes a free PDF copy of this book along with other exclusive benefits. Check the *Free Benefits with Your Book* section in the Preface to unlock them instantly and maximize your learning experience.

The evolving landscape of cyber threats

In the dynamic world of cyberspace, change is the only constant. Cyber threats continue evolving, becoming more sophisticated and diverse each year. At its core, cybersecurity is protecting systems, networks, and information from these digital threats. The **National Institute of Standards and Technology (NIST)** defines cybersecurity as *"the ability to protect or defend the use of cyberspace from cyberattacks."*

This encompasses not only the prevention of attacks but also the detection and response to malicious activities designed to access, alter, or destroy sensitive data, or disrupt essential services. To stay ahead of these threats, organizations must build their defenses on robust security frameworks, such as those developed by NIST, that adapt to the growing complexity of today's threat landscape.

The evolution of cyber threats reveals the relentless ingenuity of attackers who exploit vulnerabilities in our digital world. From the early days of simple computer viruses to the complex and persistent attacks we face today, cyber threats have reached unprecedented levels of sophistication. This constant progression challenges defenders to evolve their strategies and tools continuously, ensuring that security measures can withstand the ever-growing array of threats.

The nascent days of the internet were marked by the emergence of computer viruses and worms. These early threats were relatively straightforward in their execution but represented the first signs of the looming digital dangers. Viruses such as the infamous **ILOVEYOU** (aka Lovebug) and the **Melissa** worm wreaked havoc by spreading through email attachments, infecting computers, and causing damage to data and systems.

While these early threats garnered attention for their disruptive capabilities, they were rudimentary compared to what would follow. At this stage, cyberattacks were often driven by curiosity or mischief rather than organized criminal or nation-state activities.

As the internet matured, so did the motivations behind cyberattacks. The rise of **hacktivism** saw politically motivated groups using digital tools to promote their causes and engage in acts of online civil disobedience. Groups such as *Anonymous* and *LulzSec* gained notoriety for their high-profile attacks on government websites, corporations, and institutions they deemed as adversaries.

Hacktivism represented a new dimension of cyber threats, where ideology and social activism were driving forces behind attacks. It showcased the potential of the digital realm as a platform for political and social change.

With the proliferation of e-commerce and online financial transactions, cybercriminals found new avenues for financial gain. The digital underground became a thriving marketplace for stolen credit card data, personal information, and illicit goods. Cybercrime syndicates operated with a level of sophistication that rivaled traditional criminal organizations.

The diversity of cybercrime expanded, encompassing a wide range of activities such as identity theft, ransomware attacks, and the sale of illicit drugs and weapons on the dark web. The financial incentives for cybercriminals were substantial, driving them to develop advanced tactics and techniques to bypass security measures.

One of the most significant developments in the evolution of cyber threats has been the emergence of state-sponsored cyber espionage. Nation-states recognized the potential of cyberattacks to gain a strategic advantage in the global arena. They began investing heavily in developing cyber capabilities, creating **advanced persistent threats (APTs)**.

APTs are characterized by their stealthy and persistent nature. State-sponsored actors use APTs to infiltrate target systems, remain undetected for extended periods, and steal sensitive information. Notable examples include the Chinese hacking group *APT1*, implicated in cyber espionage against various industries, and the Russian group *APT29*, accused of interfering in foreign elections. In recent years, ransomware has emerged as one of the most pervasive and financially lucrative cyber threats.

Ransomware attacks involve encrypting a victim's data and demanding a ransom for its decryption. The attackers often threaten to permanently delete or publish the data if the ransom is not paid. What sets ransomware apart is its ability to cause immediate and tangible harm. Hospitals, municipalities, and businesses have fallen victim to ransomware attacks, leading to disrupted operations, financial losses, and even risks to public safety. Cybercriminals have honed their tactics, using phishing emails and exploiting vulnerabilities to gain initial access to systems.

The evolution of cyber threats has been characterized by a relentless march toward greater sophistication and diversity. Threat actors now employ advanced techniques such as zero-day exploits, social engineering, and supply chain attacks. They target not only traditional computer systems but also mobile devices, IoT devices, and cloud infrastructure. The motivations behind cyber threats have diversified. While financial gain remains a primary driver, state-sponsored espionage, hacktivism, and even ideological motives have gained prominence. The global interconnectedness of the digital world means that an attack on one entity can have far-reaching consequences, affecting industries, nations, and individuals.

The significance of cybersecurity

In the modern world, where every aspect of our lives is intertwined with digital technology, cybersecurity is not merely a buzzword or an afterthought but the foundation upon which our digital existence rests. The significance of cybersecurity extends far beyond the realm of technology; it is a critical element that impacts individuals, organizations, and entire nations, shaping the landscape of our interconnected world. Here, we explore why cybersecurity is not just critical but crucial in the modern era.

Personal privacy is under constant threat. Individuals rely on digital platforms for communication, financial transactions, healthcare management, and more. This reliance necessitates the exchange of sensitive information, from personal identification details to medical records and financial data.

For example, the 2017 Equifax data breach exposed the personal information of nearly 147 million Americans. The breach had far-reaching consequences, leading to identity theft and financial fraud cases, emphasizing the dire need for robust cybersecurity measures to protect sensitive data.

For organizations, downtime caused by an attack can result in financial losses, reputational damage, and even legal liabilities. In May 2017, the WannaCry ransomware attack infected over 200,000 computers across 150 countries, disrupting operations in healthcare, manufacturing, and logistics. The attack illustrated how unprepared organizations faced devastating consequences, underlining the importance of cybersecurity for business continuity.

Cybersecurity measures protect critical business systems, data, and operations. They shield against disruptive events such as ransomware attacks, DDoS attacks, and data breaches. Ensuring business continuity is not just about profitability but also about safeguarding jobs, investments, and the trust of customers and stakeholders. Organizations invest heavily in research and development to create cutting-edge products, services, and technologies. Protecting these innovations is essential for maintaining a competitive edge in the global market.

Cybersecurity plays a critical role in protecting intellectual property and national security. Cyberattacks such as corporate espionage and IP theft can undermine years of research, giving competitors unfair advantages and stifling innovation. Notably, the 2014 indictment of Chinese military members for hacking American companies highlighted the dangers of cyber espionage. On a larger scale, cybersecurity is crucial for safeguarding national infrastructure. Digital systems manage critical services such as defense, power grids, and transportation, making them vulnerable to catastrophic consequences in a cyberattack. The 2015 cyberattack on Ukraine's power grid, which left 200,000 people without electricity, underscores the potential severity of such breaches, particularly in the context of state-sponsored cyber warfare.

In the global economy, cybersecurity ensures the integrity of cross-border trade, financial transactions, and supply chains, which rely on digital technologies. Cyberattacks, such as the 2017 NotPetya ransomware attack, disrupted multinational companies, resulting in significant financial losses and supply chain chaos, illustrating how deeply the global economy is interconnected with cybersecurity. On an individual level, cybersecurity fosters trust in digital systems, encouraging engagement in digital services. However, breaches can erode this trust, as seen in the aftermath of the 2013 Target data breach, where the company suffered financially due to a loss of consumer confidence. Beyond economic concerns, cybersecurity also involves social and ethical considerations that influence how we interact with digital platforms.

Ensuring the security and privacy of digital communications is not just a matter of convenience but also human rights. Access to secure digital communication is a fundamental aspect of modern life, and cybersecurity protects this access. Additionally, the ethical implications of cyberattacks, particularly state-sponsored activities, are a subject of international debate. Ethical considerations in cybersecurity are brought to the forefront by the Stuxnet worm. This state-sponsored cyberweapon, allegedly developed by the U.S. and Israel, targeted Iran's nuclear program. Stuxnet raised ethical questions about using cyberattacks as a tool of international policy, underscoring the need for responsible behavior in the digital spectrum. The significance of cybersecurity extends to promoting responsible and ethical behavior in the digital realm, safeguarding human rights, and upholding international norms. Human behavior often plays a pivotal role. A report by Verizon highlights that 85% of data breaches involve human elements such as phishing and social engineering attacks. This statistic shows the importance of cybersecurity awareness and education for individuals navigating the digital landscape. Internationally, the European Union's **General Data Protection Regulation (GDPR)** is a prime example of how governments take cybersecurity seriously. GDPR, which took effect in 2018, places stringent requirements on organizations handling personal data, with severe financial penalties for non-compliance. It reflects a growing global recognition of the importance of data protection and cybersecurity. In conclusion, the significance of cybersecurity in the modern world is not a hypothetical concept but a defining challenge of our time. Real-world incidents, escalating financial costs, national security imperatives, human vulnerabilities, and international regulations all emphasize the critical role of cybersecurity in safeguarding our digital existence.

Types of threat actor and their motives

Cyberattackers come in various forms, depending on their skills, motivations, and the methods they use to exploit cybersecurity vulnerabilities. Understanding the different types of cyberattackers is crucial for implementing effective security measures. Here are some common types of threat actors:

- **Hacktivists**: These are cyberattackers who are driven by ideological causes or seek to bring about social or political change. They often target organizations or individuals they perceive as acting against their principles. Their activities may include website defacement, launching DoS attacks, or stealing and leaking sensitive information.

- **Script kiddies**: This group is often amateurs who lack sophisticated hacking skills. They use readily available tools and scripts to launch attacks, not necessarily understanding the underlying technology. Their motives are often trivial, driven by a desire for recognition or personal amusement rather than monetary or political reasons.

- **Cybercriminals**: These attackers are part of organized crime groups or operate individually. They are motivated by financial gain and use a variety of methods, including malware, phishing, and other forms of fraud, to steal data, money, or other valuable information.

- **State-sponsored attackers**: These are hackers employed by governments to conduct cyber espionage, sabotage, or warfare activities. Their targets can be other nations, key infrastructure, political groups, or individuals. They are generally well-funded and very skilled, capable of launching sophisticated and persistent attacks.

- **Insiders**: These attackers are individuals within an organization who misuse their access rights to intentionally harm their employer. They might be driven by financial incentives, grievances, or coercion by external forces. Insider threats are dangerous because these individuals already have knowledge of and access to internal systems.

- **Lone wolf hackers**: These are individuals who operate independently and may have various motivations, including financial gain, challenge, or personal beliefs. They might be highly skilled and choose targets based on opportunity or personal interest.

- **Cyber terrorists**: These attackers seek to create fear and chaos by disrupting digital systems. They may be affiliated with terrorist groups and engage in attacks meant to cause widespread panic, harm, or financial disruption, often pursuing political, religious, or ideological goals.

- **APTs**: These groups are usually state-sponsored or may act with state-like capabilities. They pursue long-term campaigns and are equipped to remain undetected within a network for extended periods. Their goals typically involve espionage, disruption, or theft of sensitive data.

Having identified the various types of cyberattackers and their motives, we now clearly understand the threat landscape. This knowledge sets the stage for delving into the cyber kill chain, a framework designed to detail the steps attackers take to achieve their objectives. Understanding this sequential process can better anticipate, detect, and disrupt cyberattacks at various stages. In the next section, we will introduce the concept of the cyber kill chain, outlining its stages and explaining how it can be employed to bolster our cyber defense strategies.

Introduction to the Cyber Kill Chain concept

Enter the **Cyber Kill Chain**, a powerful model that unveils the inner workings of cyberattacks, equipping defenders with the knowledge needed to thwart them effectively.

The model, initially developed by Lockheed Martin, serves as a roadmap for comprehending and countering cyber threats. The CKC model borrows its name from military terminology, where a **kill chain** represents the sequence of events leading to the destruction of a target. Its primary purpose is to break down the anatomy of a cyberattack into distinct stages, each with its own objectives and characteristics. By doing so, it empowers cybersecurity professionals to anticipate, prepare for, and mitigate cyber threats at various points along the attack continuum.

The CKC comprises seven interlinked stages, each representing a critical phase in the life cycle of a cyberattack. These stages are as follows:

1. **Reconnaissance**: This initial phase involves gathering information about the target. Attackers seek to identify vulnerabilities, potential entry points, and weaknesses within the target's infrastructure. The 2015 cyberattack on the U.S. **Office of Personnel Management** (**OPM**) serves as a stark example. Attackers, believed to be state-sponsored, conducted extensive reconnaissance to gather sensitive personal information of millions of federal employees, highlighting the critical importance of this initial phase. In 2018, Marriott disclosed a massive data breach affecting over 500 million guests' records. The breach involved hackers exploiting vulnerabilities over several years, highlighting the persistence of APTs in today's cybersecurity landscape. The attack resulted in the exposure of sensitive personal information, including passport numbers, which demonstrates how attackers can remain undetected within systems for long periods, further emphasizing the need for early detection and response measures in the CKC.

2. **Weaponization**: Once armed with intelligence about the target, attackers prepare malicious payloads, such as malware or exploit kits. These are weaponized to facilitate the attack. The Stuxnet worm, discovered in 2010, exemplifies the *Weaponization* stage. It was a highly sophisticated cyberweapon believed to be developed by nation-states. Stuxnet targeted industrial control systems, underscoring the ability of attackers to craft intricate digital weapons.

3. **Delivery**: Attackers deliver the weaponized payload to the target system. Common delivery methods include phishing emails, malicious links, or compromised websites. The 2016 phishing campaign against John Podesta, chairperson of Hillary Clinton's presidential campaign, demonstrated the potency of delivery. Podesta fell victim to a spearphishing email, leading to the breach of his email account—a breach with far-reaching consequences.

4. **Exploitation**: In this stage, attackers exploit vulnerabilities within the target system to gain initial access. This often involves leveraging software vulnerabilities or social engineering techniques. The WannaCry ransomware attack in 2017 exploited a known vulnerability in Windows systems. This cyber epidemic affected hundreds of thousands of computers worldwide, underscoring the potential damage caused by exploitation.

5. **Installation**: After gaining a foothold, attackers install malware or establish a persistent presence within the compromised system. This stage ensures they maintain control even if detected initially. For instance, the **SolarWinds breach** of 2020 demonstrated how attackers could compromise software supply chains, affecting thousands of organizations, including critical government agencies. Similarly, the resurgence of **Emotet** in 2021, after a coordinated international takedown, highlights the persistent threat posed by banking malware, evolving from earlier trojans such as Zeus. In addition, the latest various data breach reports reveal an alarming rise in ransomware attacks, underscoring the growing prominence of these destructive threats.

6. **Command and Control**: Attackers establish communication channels with the compromised system, allowing them to control it remotely. This stage enables them to carry out malicious actions and exfiltrate data. The Zeus banking Trojan is an illustrative example of the *Command and Control* stage. It enabled cybercriminals to remotely control infected computers and steal sensitive financial information.

7. **Actions on Objectives**: The final stage involves the attackers' actual objectives, which could include data exfiltration, system disruption, or other malicious activities aligned with their goals. The breach of Sony Pictures Entertainment in 2014 serves as a stark illustration. Attackers, allegedly linked to North Korea, unleashed destructive malware and exposed sensitive data in an attempt to stifle the release of a film they deemed offensive.

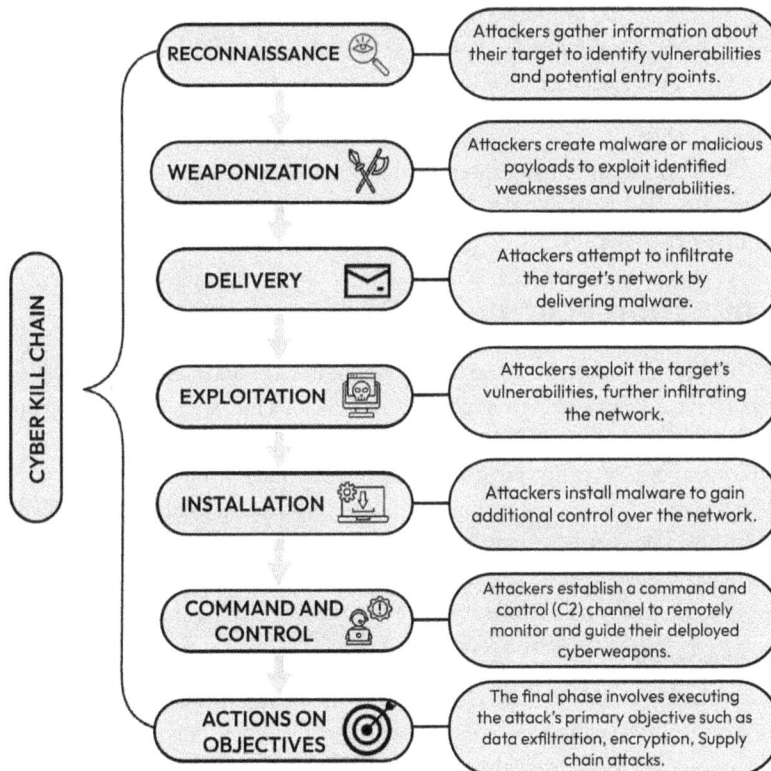

Figure 1.2 – Cyber kill chain

Understanding the cyber kill chain is akin to studying an adversary's playbook. By breaking down the attack into these discrete stages, cybersecurity professionals gain insights into how attackers think, plan, and execute their strategies. This knowledge provides several crucial advantages, such as the following:

- **Early detection**: Recognizing the signs of an attack during the initial stages, such as *Reconnaissance* and *Weaponization*, allows defenders to respond proactively, preventing the attack from progressing further.

- **Threat intelligence**: Analyzing the **tactics, techniques, and procedures (TTPs)** used in each stage enables organizations to develop threat intelligence. This intelligence can be shared with the broader cybersecurity community to enhance collective defense.

- **Tailored defense**: Armed with an understanding of the cyber kill chain, organizations can develop tailored defense strategies. This involves implementing security measures and controls specific to each stage, making it harder for attackers to advance.

- **Incident response**: In the unfortunate event of a successful breach, the cyber kill chain provides a framework for incident response. It helps identify where and how the attack occurred, aiding in recovery and preventing future incidents.

The cyber kill chain is not a static model but an evolving one. As cyber threats continue to advance, so does the model itself, adapting to adversaries' changing tactics and technologies. The concept is not a theoretical construct but a tangible framework that provides a lens through which we can dissect, understand, and counter cyberattacks.

Real-world incidents underscore its significance, revealing the intricate dance between attackers and defenders. As we delve deeper into the chapters of this book, we will explore each stage of the cyber kill chain, learning from real-world incidents and leveraging this knowledge to fortify our defenses in an ever-evolving cyber landscape.

The state of cybersecurity in 2024

Cybersecurity in 2024 presents a challenging landscape as threats continue to evolve, driven by both technological innovation and increasingly sophisticated attack techniques. The rise of **crimeware-as-a-service (CRaaS)** platforms has made it easier for attackers with minimal technical skills to execute large-scale cyberattacks. Additionally, ransomware strategies have shifted from simply encrypting data to focusing on extortion, as many organizations have implemented robust backup and recovery solutions. Attackers now steal sensitive data and threaten to release it unless ransoms are paid, further complicating the defense landscape. This dynamic environment demands that organizations adopt more proactive and resilient security strategies to counter the escalating threat levels.

This evolving threat landscape is closely aligned with the stages of the cyber kill chain. In the *Reconnaissance* and *Weaponization* phases, attackers increasingly automate their efforts, using CRaaS platforms to identify vulnerabilities and develop attack vectors at scale. To combat these threats, defenders must enhance their detection capabilities, utilizing advanced tools such as **artificial intelligence (AI)** and machine learning to anticipate potential attacks and respond swiftly. The shift in ransomware tactics underscores the need for heightened vigilance during the *Delivery* and *Exploitation* stages, where attackers seek to penetrate an organization's defenses. Addressing these early-stage threats is crucial, especially as attackers exploit social engineering and supply chain vulnerabilities to gain entry.

In response to these challenges, the adoption of Zero Trust architecture has gained momentum as a core defense strategy in 2024. Zero Trust principles, such as continuous verification and least-privilege access, are designed to mitigate threats during the later stages of the cyber kill chain, particularly in the *Command and Control (C2)* and *Actions on Objectives* phases. Organizations can significantly reduce the potential damage of a breach by limiting lateral movement within a network and ensuring that users and systems only have access to the data they need. This approach is critical as attackers increasingly focus on data theft and extortion rather than merely disrupting operations. Implementing strong access controls and segmenting networks can disrupt the attack chain and prevent adversaries from achieving their objectives.

The cost of a data breach

The escalating costs and impact of data breaches are closely linked to the stages of the cyber kill chain, offering valuable insights into how attackers progress and where defenders can intervene to mitigate damage. In the *Reconnaissance* and *Weaponization* phases, attackers exploit vulnerabilities such as unmanaged data (shadow data) and compromised credentials. In 2024, 35% of breaches involved shadow data, leading to a 16% increase in costs. The prolonged detection times, sometimes exceeding 250 days for phishing or social engineering, underscore the urgent need for proactive measures in the early stages of the cyber kill chain, particularly in the *Reconnaissance* phase.

AI and automation are crucial in disrupting the *Delivery* and *Exploitation* phases of the cyber kill chain. Organizations utilizing these technologies reduced breach costs by an average of $2.2 million, as AI tools assist in identifying malicious payloads and suspicious behavior faster. Without AI, breach costs averaged $5.72 million, but with extensive AI use, this number dropped to $3.84 million. These technologies allow defenders to detect threats before attackers can fully exploit systems, helping to break the chain during exploitation.

In the *Installation* phase, attackers often use shadow data or unmanaged environments to plant malware or establish backdoors. This results in longer breach life cycles and higher costs. On average, breaches involving shadow data cost $5.27 million. As attackers take advantage of disorganized data management, organizations must improve their visibility across environments to disrupt the kill chain during installation and reduce the financial impact.

During the *C2* phase, AI tools are essential for detecting unusual network activity and stopping attackers from maintaining control. Involving AI and automation allowed organizations to reduce the time to identify and contain a breach by almost 100 days. This quicker response disrupts the *C2* stage and prevents the attack's completion, helping include operational and financial damage.

In the *Actions on Objectives* phase, attackers aim to complete their mission—whether it be data exfiltration or disruption. Malicious insider attacks cost organizations an average of $4.99 million, making them one of the most expensive breach types. By integrating AI across all phases of the cyber kill chain, especially during prevention, detection, and response, organizations can reduce the time attackers have to achieve their goals and the overall financial impact of a breach.

Summary

The **Cyber Kill Chain** framework remains a fundamental tool for cybersecurity professionals, offering a detailed roadmap of the stages attackers follow during a cyberattack. By understanding and anticipating each step, from reconnaissance to execution, organizations can deploy defenses that disrupt an attack before significant damage occurs. Recent incidents, such as the **2020 SolarWinds supply chain attack** and the **2021 Colonial Pipeline ransomware attack**, underscore the importance of this framework in mitigating modern threats. The SolarWinds breach highlighted how adversaries could compromise trusted software updates, while the Colonial Pipeline attack demonstrated the real-world impact of cyberattacks on critical infrastructure. These cases exemplify the need for organizations to adopt proactive defense strategies aligned with the cyber kill chain, allowing for early detection and swift response to minimize the consequences of evolving cyber threats.

Understanding the cyber kill chain is essential for security professionals as it enables them to implement targeted defense strategies effectively. By recognizing the indicators of compromise at each stage, security teams can detect threats earlier and respond more swiftly to disrupt the attack progression. This proactive approach minimizes the impact of attacks and enhances the overall security posture by allowing continuous refinement of security measures based on observed attacker behaviors and tactics. Thus, mastery of the cyber kill chain concept equips professionals with the knowledge to react to and preempt cyber threats, significantly bolstering an organization's cybersecurity defenses.

With a concrete understanding of the cyber kill chain's foundational concepts established in this chapter, we will now delve into the first phase of this model: *Reconnaissance*. This crucial initial step involves attackers gathering valuable intelligence about their target to identify potential vulnerabilities and develop an effective strategy for exploitation. By leveraging techniques such as scanning networks, harvesting emails, and monitoring social media activity, attackers can build a comprehensive profile of their target's defenses and operational behaviors. This phase sets the stage for the subsequent steps in the cyber kill chain, highlighting the importance of proactive measures and robust security protocols to detect and mitigate reconnaissance activities before they escalate into more advanced stages of an attack. In the following chapter, we will explore these reconnaissance techniques in detail, providing insights into the methods used by cyber adversaries and the countermeasures that can be employed to thwart their efforts.

Further reading

Following are the references we used in this chapter along with some bonus reading materials for you to check out:

- *Lockheed Martin's overview of the Cyber Kill Chain® framework for identifying and preventing cyber intrusions* – `https://www.lockheedmartin.com/en-us/capabilities/cyber/cyber-kill-chain.html`

- *Wikipedia article about the 2015 cyberattack on Ukraine's power grid* – `https://en.wikipedia.org/wiki/2015_Ukraine_power_grid_hack`

- *New York Times article on the 2017 WannaCry ransomware attack that affected the UK's National Health Service* – `https://www.nytimes.com/2017/05/12/world/europe/uk-national-health-service-cyberattack.html`

- *Wikipedia article detailing the 2017 Equifax data breach that exposed personal information of millions* – `https://en.wikipedia.org/wiki/2017_Equifax_data_breach`

- *Wired article about Operation Aurora, a series of cyberattacks against major companies in 2009* – `https://www.wired.com/2010/01/operation-aurora/`

- *SANS Institute's 2024 CISO Primer on cybersecurity trends and best practices* – `https://www.sans.org/mlp/ciso-primer-2024/`

- *IBM's Cost of a Data Breach Report 2024 analyzing the financial impact of data breaches* – `https://www.ibm.com/downloads/cas/1KZ3XE9D`

- *Research paper on the evolution of Security Operations Centers from reactive to proactive strategies* – `https://vipublisher.com/index.php/vij/article/view/589`

- *Study on leveraging artificial intelligence to enhance security operations while balancing efficiency and human oversight* – `https://vipublisher.com/index.php/vij/article/view/588`

2

Reconnaissance – The Initial Breach Plan

The *Reconnaissance (Recon)* phase, often considered the initial stage of the Cyber Kill Chain, is a critical step in any cyberattack. In this phase, attackers focus on gathering public information about their targets, which can include individuals, organizations, or even entire nations. The primary goal here is to lay the foundation for a successful attack by gaining valuable insights into the target's vulnerabilities, systems, and potential entry points.

In this chapter, you will gain a good understanding of the importance of this phase from the perspective of both a cybersecurity professional and a cyber adversary such as nation-state actors, hacktivists, cyber criminals, and so on. Recon can be broadly categorized into two main types: **passive** and **active**. It's essential to understand the distinction between the two approaches, as they play a critical role in the attacker's strategy.

In this chapter, you'll learn about the following topics:

- Significance of Recon
- Passive Recon
- Active Recon
- Real-life incidents
- Attack surface management
- Techniques of detecting Recon activities
- Practical tools for Recon

Let's dive in!

The significance of the Recon phase

The Recon stage is critical for the success of the subsequent phases of the cyber kill chain and can shape the effectiveness of the attack. Here's why the Recon stage is so critical:

- **Information gathering**: In this phase, the attacker gathers as much information as possible about their target. This can include network information, domain details, employee information, and more. The quantity and the quality of the information collected at this stage can determine how well the attacker can tailor the sequence of the attack.

- **Target identification**: The attacker identifies the potential vulnerability within the target organization. They generally look for weak spots in the network defense, such as outdated systems, unpatched endpoints, or unpatched software, which can be exploited in the attack chain.

- **Social engineering attacks**: One of the goals of this stage is to prepare for the next step. Attackers may use the information gathered to craft convincing phishing emails or other social engineering tactics such as phishing, pretexting, baiting, quid pro quo, tailgating/piggybacking, vishing, spear phishing, watering hole attacks, and so on. These often seem very credible because they are based on the actual data.

- **Strategic advantage**: Understanding the target's business operations, hierarchy, and schedule can provide strategic advantages. For instance, knowing when the company is about to make a significant announcement can help the attackers to maximize impact for the following stages.

Understanding the significance of the Recon phase in cybersecurity lays the groundwork for delving into the specifics of the passive Recon stage within the broader context of the Recon phase in the cyber kill chain.

Passive Recon

Passive Recon is a fundamental stage in the cyber kill chain, representing the initial step in an attacker's quest for information about a target. It's characterized by data collection without direct interaction with the target. This methodological approach provides cyber adversaries with a wealth of information that can be pivotal in planning and executing a successful attack. In this phase, attackers rely on publicly available data sources, such as websites, social media profiles, **Domain Name System** (**DNS**) records, and WHOIS information. They also employ search engines and open source intelligence tools to compile a comprehensive profile of the target. One of the key advantages of passive Recon is its subtlety; it operates discreetly and is less likely to trigger security alerts or raise suspicion.

The following are the sources of passive Recon:

- **Search engines**: Search engines are invaluable tools for passive Recon, allowing attackers to uncover sensitive information with minimal effort. By using advanced search operators, often called **Dorking** (e.g., Google Dorking), cyber adversaries can sift through indexed content to find exposed email addresses, unprotected network configurations, and potentially vulnerable systems. Search engines such as Google, Bing, and Yahoo index vast amounts of publicly accessible data, making them a treasure trove of information for those with malicious intent.

- **Social media**: Social media platforms are treasure troves of information for attackers. Individuals associated with the target, whether they are employees or otherwise, often inadvertently share personal details, affiliations, and even potential passwords. In the hands of a skilled attacker, this publicly disclosed information can be leveraged to breach the target's defenses.

- **Job listings and forums**: Online job listings and forums such as LinkedIn, Indeed, Monster, Dice, and so on can be surprising sources of valuable information. Job listings might disclose the technologies and systems used by the target organization. Employees seeking assistance or discussing their work on forums may unknowingly give attackers valuable insights.

- **DNS records**: DNS records are a goldmine for passive Recon. Attackers can glean vital insights into the target's infrastructure, including subdomains and IP addresses. This detailed understanding enables them to map out the network's architecture, which is crucial for a successful attack.

- **WHOIS**: The full form of WHOIS is "*Who Is.*" It is a query and response protocol used for querying databases that store the registered users or assignees of an internet resource, such as a domain name, an IP address block, or an autonomous system. The WHOIS database contains information such as the registrant's name, contact information, and the registration and expiration dates of the domain.

Transitioning from passive Recon, which focuses on gathering information without direct interaction, we now move to active Recon, where attackers engage more directly with the target to obtain detailed and specific data.

Active Recon

Active Recon, the next phase in the Cyber Kill Chain, takes a more proactive and direct approach. Attackers actively engage with the target's systems to gather information. While this approach provides a deeper understanding of the target's infrastructure and potential vulnerabilities, it also carries a higher level of risk for the attacker, as it may leave digital footprints that could be detected by security systems.

The following techniques are associated with active Recon:

- **Port scanning**: Port scanning is a core technique in active Recon. Attackers employ port scanning tools to identify open ports on the target's systems. Open ports can indicate potential entry points for exploitation. Understanding the services and ports that are accessible is crucial for attackers in the Recon stage. By probing for open ports, attackers can identify vulnerabilities in the network's defenses.

- **Network mapping**: Network mapping, facilitated by Nmap, enables attackers to create a detailed map of the target network. This map outlines the devices and services in operation, providing attackers with a clear picture of the network's layout. With this information, attackers can identify high-value targets and potential weak points in the infrastructure.

- **Vulnerability scanning**: Vulnerability scanning is a pivotal step in active Recon. Attackers use scanning tools to identify known vulnerabilities in software and services. This process is essential, enabling attackers to pinpoint areas where the target's defenses may be lacking, as well as vulnerabilities that can be exploited.

In conclusion, passive and active Recon are pivotal stages in the cyber kill chain, wherein attackers gather critical information to identify potential vulnerabilities and plan their attacks. Passive Recon focuses on non-intrusive data collection, while active Recon involves more direct interactions with the target's systems. Both stages are crucial for threat actors to successfully progress through the Kill Chain and execute cyberattacks.

In this digital age, attackers have a wealth of tools and techniques to carry out Recon effectively. In this phase, attackers typically seek answers to critical questions:

- What are the target's assets and infrastructure?
- Who are the key personnel within the organization?
- Which software and hardware is in use?
- Are there any known vulnerabilities that can be exploited?

Having explored the various stages and techniques within the Recon phase, let's now turn our attention to real-life incidents when these methods have been employed, highlighting the practical implications and impact of effective Recon in cybersecurity breaches.

The Recon phase is critical in planning and executing cyber attacks. This phase involves passive and active Recon techniques, each playing a distinct role in gathering essential information about the target. Passive Recon allows attackers to quietly collect data without interacting with the target systems, thus minimizing the risk of detection. In contrast, active Recon involves direct interaction with the target, providing more detailed and precise information but with a higher risk of exposure.

The importance of the Recon phase cannot be emphasized enough, as the information gathered during this stage forms the bedrock for subsequent attack vectors. Understanding the target's infrastructure, vulnerabilities, and defense mechanisms empowers attackers to design more effective and tailored strategies to exploit weaknesses. By comprehending the intricacies of both passive and active Recon, defenders can proactively anticipate potential threats and develop robust security measures to thwart attackers early in the Cyber Kill Chain, instilling a sense of confidence in their ability to protect their networks.

Overall, mastering the Recon phase is crucial for attackers seeking to infiltrate systems and defenders aiming to protect their networks. Effective Recon can distinguish between a successful breach and a failed attempt, underscoring its pivotal role in cybersecurity.

Real-life incidents – Recon in action

In order to demonstrate the practical impact and implementation of Recon techniques, we will explore real-life incidents when these methods were crucial. The cases of Stuxnet, the Target data breach, and Operation Aurora offer compelling examples of Recon in action. They illustrate how thorough intelligence gathering can result in significant cybersecurity breaches and underscore the need for strong defensive measures.

Stuxnet attack (2010)

The **Stuxnet worm**, believed to be a joint operation by the U.S. and Israel, is a prime example of the importance of Recon. Researchers at Symantec detailed the attack in their *Stuxnet 0.5 – The Missing Link* report, which thoroughly explains how the attackers conducted Recon. The Stuxnet attack of 2010 is a pivotal moment in the history of cyber warfare. This operation showcased a remarkable blend of cyber espionage and sabotage tactics, with its primary target being Iran's nuclear facilities, specifically the Natanz uranium enrichment plant.

The attackers extensively studied Natanz's **Supervisory Control and Data Acquisition (SCADA)** systems, particularly Siemens Step7 software. Symantec's *Stuxnet Dossier* provides insights into Stuxnet's targeting of SCADA systems. Understanding the behavior of centrifuges and critical systems within Natanz was crucial. This information informed the tailored attack strategy. The attackers utilized a previously unknown software flaw that had not been patched (in other words, ZeroDay), leaving it open to exploitation. This is also referred to as using zero-day vulnerabilities to gain access, exploiting flaws that are unknown to software vendors. USB drives were chosen as a transmission vector, taking advantage of the contractors' use of them to transfer data into Natanz's isolated network. This type of attack is also known as a **Candy Drop** attack. Ralph Langner's *Stuxnet: Dissecting a Cyberwarfare Weapon* details Stuxnet's propagation methods. Stuxnet was designed to only execute on Windows-based systems with specific configurations, ensuring precision in targeting.

The Stuxnet attack's Recon phase set the stage for its unprecedented impact. Once deployed, the malware disrupted Iran's uranium enrichment operations. Thousands of centrifuges were damaged or destroyed, significantly delaying Iran's nuclear program. Stuxnet's intricacy and sophistication continue to astound cybersecurity experts and academics alike. It serves as a benchmark for cyber espionage and sabotage operations.

Target data breach (2013)

Brian Krebs extensively covered the Target breach on his blog, in the *Target Hackers Broke in Via HVAC Company* post. The attackers conducted Recon by compromising Fazio Mechanical Services, a third-party **Heating, Ventilation, and Air Conditioning (HVAC)** contractor for Target. The Target data breach remains a landmark case in the world of cybersecurity, serving as a stark reminder of the vulnerabilities that retailers face in the digital age.

The attackers behind the Target data breach executed a calculated plan that commenced with a comprehensive Recon phase:

1. The Recon phase began when the cybercriminals compromised a third-party HVAC contractor that had access to Target's network.

2. The attackers mapped Target's network architecture, identifying critical systems and vulnerabilities.

3. They zeroed in on Target's **Point-of-Sale (POS)** systems, which were critical to their plan. They identified weaknesses in these systems.

4. The Recon phase also involved crafting convincing spear-phishing emails. These emails targeted specific employees within Target. The Recon phase laid the groundwork for the development of malware tailored to Target's environment.

S   **Support**   Date:
         support@securebank.com   **19-06-2024 14:24:55**

Subject: Immediate Action Required: Verify Your Account

Dear Recipient,
We have detected unusual activity on your account that requires immediate attention. For your security, we need you to verify your account information to prevent any unauthorized access.
To verify your account, please click the link below:
Verify Your Account
Failure to verify your account within 24 hours will result in temporary suspension of your account. We apologize for any inconvenience this may cause, but your security is our top priority.
Important:
Do not share your login details with anyone.
If you did not initiate this request, please contact our support team immediately at support@securebank.com.
Thank you for your prompt attention to this matter.
Sincerely,
Customer Support Team
SecureBank

Figure 2.1 – Spear-phishing email

The culmination of the Recon phase was the infiltration of Target's network. Once inside, the attackers unleashed malware that harvested customer credit card data. This breach compromised millions of payment card records and severely damaged Target's reputation.

Operation Aurora (2009)

Operation Aurora, which emerged in 2009, is a prime example of state-sponsored cyber espionage with a significant Recon component.

The operation's name, *Aurora*, is derived from the Aurora Flight Sciences company, one of its early victims. Operation Aurora came to light when Google publicly disclosed that it had fallen victim to a cyber-attack originating from China. This revelation shed light on a broader campaign impacting several prominent companies.

Operation Aurora had the following clear objectives:

- **Corporate espionage**: The attackers sought to gain unauthorized access to the networks of targeted corporations, enabling them to steal valuable corporate data and intellectual property.

- **Exfiltration of sensitive information**: Beyond gaining access, the attackers aimed to exfiltrate sensitive information, such as source code, strategic plans, and proprietary technology.

Operation Aurora cast a wide net, targeting numerous corporations, including Google, Adobe Systems, Juniper Networks, and many more. The attackers exploited zero-day vulnerabilities to compromise systems and maintain a presence within the networks.

McAfee's report, *Global Energy Cyberattacks: 'Night Dragon'*, delves into the Recon phase of the Operation Aurora attacks.

The Recon phase was initiated by identifying high-profile corporations as targets. This included major technology companies. The attackers crafted highly convincing spear-phishing emails. These were strategically sent to specific employees within the targeted companies. Recon identified vulnerabilities in commonly used software. These vulnerabilities were then exploited with zero-day exploits. *Operation Aurora: A Case Study in Responding to Advanced Threats* by McAfee (please refer to the *Further reading* section for more information) discusses the exploitation of zero-day vulnerabilities. Recon determined what data to target and exfiltrate. Intellectual property and sensitive corporate information was prioritized. The attackers sought to maintain persistent access to the compromised networks, ensuring prolonged data access.

Operation Aurora had a profound impact on the targeted corporations, resulting in data breaches and intellectual property theft. It also raised awareness of the capabilities of state-sponsored cyber actors. This operation remains a case study of how Recon-driven attacks can compromise even well-defended organizations. The comprehensive Recon phase was instrumental in the attackers' ability to infiltrate and exfiltrate valuable data.

Techniques for detecting Recon activities

Understanding and identifying Recon activities is crucial for maintaining robust cybersecurity. Here, we explore practical techniques for detecting these preliminary stages of cyber attacks to better protect your network and sensitive information:

- **Network monitoring and analysis**: Recon activities often begin with the scanning of a target's network to identify vulnerabilities and potential entry points. Effective network monitoring and analysis are essential for early detection.

- **Network Intrusion Detection Systems (NIDS)**: NIDS play a crucial role in proactive network security, serving as the linchpin of defense against potential cyber threats. NIDS tools such as Snort and Suricata provide continuous, real-time analysis of network traffic, scrutinizing data for suspicious patterns and signatures associated with Recon activities. These systems are vital for early threat detection and empower rapid response when necessary.

  NIDS operates by meticulously inspecting both inbound and outbound network traffic. When it identifies anomalies that match known threat patterns, it triggers immediate alerts. These alerts are invaluable early warning signals for security teams, providing a clear indication of potential Recon activities.

  What sets NIDS apart is its proactive approach to identifying Recon activities before they escalate into full-blown attacks. Whether attackers are probing for vulnerabilities, attempting to map the network's assets, or engaging in other Recon behaviors, NIDS provides a crucial advantage in detecting and mitigating these threats. A study conducted by the **National Institute of Standards and Technology (NIST)** found that NIDS could detect up to 95% of Recon activities, significantly enhancing network security. This statistic underscores the importance of incorporating NIDS into any comprehensive security strategy.

- **Packet Capture and Analysis (PCAP)**: In addition to NIDS, PCAP forms another indispensable component of effective network monitoring. Tools such as Wireshark provide security teams with the ability to capture and scrutinize network packets, allowing for a deeper level of analysis.

  Packet analysis goes beyond simply detecting Recon activities; it can also reveal potential vulnerabilities and network misconfigurations. For example, if an attacker attempts a port scan, Wireshark captures these activities. This information empowers security teams to proactively address these vulnerable entry points before any potential exploitation can occur.

  Furthermore, a report by the SANS Institute highlights the significance of packet analysis in network security. It states that packet analysis can uncover hidden threats and vulnerabilities that may remain undetected by other security measures, making it a critical tool for Recon detection.

- **Honeypots and honeytokens**: Another effective strategy for identifying Recon activities is the deployment of honeypots and honeytokens. These are decoy systems or data deliberately designed to lure potential attackers. When accessed or manipulated, they trigger alerts, serving as clear indicators of Recon.

  Honeypots, which mimic legitimate network assets, and honeytokens, tantalizing pieces of data meant to attract attackers, share the same fundamental purpose: to detect malicious intent. Tools such as Cowrie and Dionaea are commonly used for honeypot deployment. These mimic services, servers, or endpoints that could pique an attacker's interest. If an attacker attempts to exploit these honeypots, alerts are triggered, highlighting potential threats. Honeytokens, on the other hand, are often embedded within the network, typically within documents or database entries. When they are accessed or tampered with, they indicate unauthorized access and potential Recon.

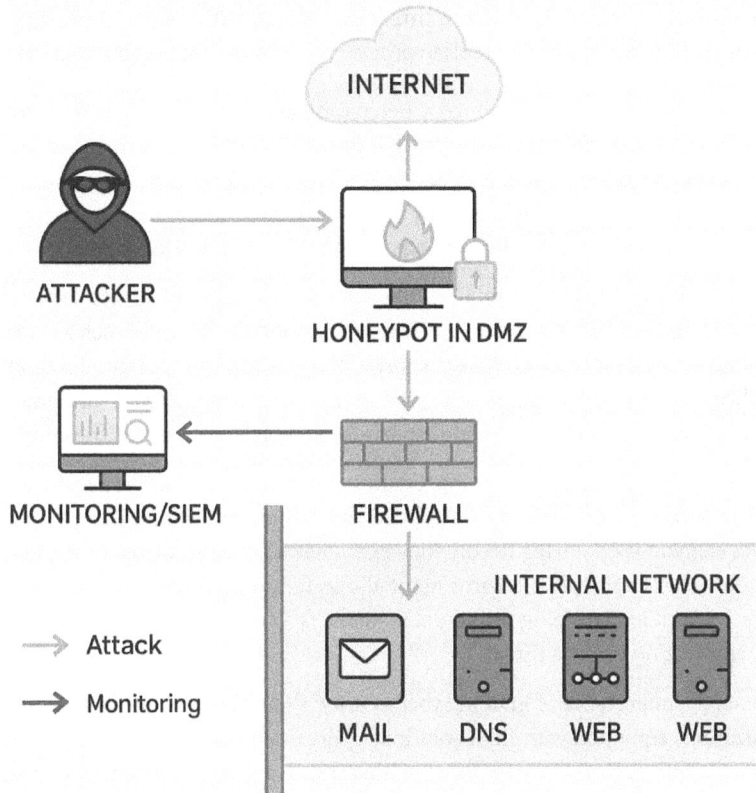

Figure 2.2 – Honeypot

Network monitoring and analysis are indispensable for early detection during the Recon stage, a critical precursor to cyberattacks. NIDS, PCAP, and the use of honeypots and honeytokens form a robust defense, enabling organizations to actively detect and respond to Recon activities.

- **Endpoint Detection and Response (EDR):** EDR is a cornerstone of modern cybersecurity, particularly during the Recon stage of the cyber kill chain. Recon often involves probing endpoint devices, such as computers and servers, for vulnerabilities or sensitive information. EDR solutions play a pivotal role in identifying and mitigating these threats at the endpoint level, making them an indispensable asset in an organization's defense strategy. EDR tools stand out with several critical features that enhance their effectiveness against sophisticated threats. One such feature is behavioral analysis, which leverages advanced monitoring of processes, file changes, and system calls to detect anomalies in endpoint behavior, excelling particularly in identifying Recon activities that evade traditional detection methods. Endpoint Visibility is another key feature, whereby endpoint agents provide real-time monitoring of device activities such as login attempts, file modifications, and registry changes. This visibility is invaluable for swiftly identifying and responding to potential threats such as brute force login attempts or system probing. Additionally, sandboxing capabilities within EDR tools allow for the execution and analysis of suspicious files or applications in an isolated environment, ensuring in-depth threat assessment without compromising the actual system. These features empower EDR tools to provide robust protection against many cyber threats:

- **Behavioral analysis**: EDR tools employ behavioral analysis, a vital approach for detecting anomalies in endpoint behavior. This analysis encompasses the monitoring of processes, file changes, and system calls. Notably, it excels at identifying Recon activities, such as unusual process execution or atypical file access. By focusing on behavior rather than known signatures, EDR tools excel in uncovering Recon activities that evade traditional detection methods.

- **Endpoint visibility**: This is central to EDR solutions are endpoint agents, offering real-time visibility into endpoint device activities. These agents monitor activities such as login attempts, file modifications, and registry changes, providing insight into potential Recon activities, such as brute force login attempts or probes of the system's registry for vulnerabilities. This visibility is invaluable, enabling swift responses to emerging threats at the endpoint level.

- **Sandboxing**: Certain EDR solutions provide sandboxing capabilities, a feature paramount in dealing with suspicious files or applications. When EDR tools identify a potentially malicious element used during Recon, they execute it within an isolated environment known as a sandbox. This controlled environment allows for an in-depth analysis of the element's behavior without jeopardizing the actual system. It's a crucial tool for determining whether a file or application poses a threat.

To underscore the importance of EDR in cybersecurity, a study by the Ponemon Institute revealed that organizations equipped with EDR solutions experienced significantly faster detection and response times with regard to cyber threats. This translated into substantial cost savings and reduced the impact of data breaches, underscoring the critical role EDR plays in enhancing an organization's cybersecurity posture.

EDR solutions stand as a critical component of a robust cybersecurity strategy, particularly during the Recon stage of the cyber kill chain. Their application of behavioral analysis, endpoint visibility, and sandboxing capabilities empowers organizations to effectively detect and respond to Recon activities at the endpoint level, ensuring the security of their networks and sensitive data.

- **DNS monitoring and analysis**: DNS is a prime target during the Recon stage of the cyber kill chain, as attackers often exploit it for their malicious activities. Rigorous monitoring and analysis of DNS traffic are imperative for unveiling potential threats. DNS monitoring and analysis can involve the following measures:

  - **DNS sinkholes**: One proactive measure in DNS Recon detection is the creation of DNS sinkholes. These are strategically set up by security teams to redirect suspicious DNS requests to controlled servers. This redirection enables the identification of Recon attempts, particularly those involving domain lookups for known malicious domains. DNS sinkholes serve as a barrier, preventing attackers from reaching their intended malicious destinations, and instead, their activities are redirected for scrutiny.

- **Passive DNS analysis**: Another invaluable tool in the arsenal of DNS Recon detection is passive DNS analysis. Tools such as DNSTwist and PassiveTotal specialize in tracking historical DNS data. By doing so, they uncover patterns of domain registrations and DNS resolutions that may be indicative of Recon efforts. This historical perspective offers insight into the evolving tactics of attackers, as they often revisit domains and IP addresses when conducting Recon.

To bolster the significance of DNS monitoring and analysis in Recon detection, a study published in the *Journal of Cybersecurity Research* highlights the effectiveness of DNS-based threat intelligence in identifying malicious domains and activities. The study demonstrates that proactive DNS monitoring, in conjunction with the use of DNS sinkholes and passive DNS analysis, can significantly enhance an organization's security posture.

DNS is a primary focus during the Recon stage of the cyber kill chain, making it crucial to adopt comprehensive monitoring and analysis strategies. DNS sinkholes and passive DNS analysis tools provide a robust defense against Recon efforts involving DNS, offering organizations a proactive means of detecting and mitigating threats.

Having explored various techniques for detecting Recon activities, it is essential to delve into the role of threat intelligence feeds in bolstering these detection capabilities.

CTI and threat intelligence feeds

Cyber Threat Intelligence (**CTI**) involves collecting and analyzing information about current and potential cyber threats, enabling organizations to understand and anticipate malicious activities. CTI empowers cybersecurity teams to make informed decisions and proactively defend against evolving threats.

Leveraging threat intelligence feeds is a proactive and effective strategy. These feeds serve as invaluable sources of up-to-date information on known threats and **Indicators of Compromise** (**IoCs**). By incorporating threat intelligence feeds into their defense strategies, organizations gain a critical edge in identifying and responding to Recon activities. Following are the key areas where these feeds integrate and play a vital role:

- **Integration with Security Information and Event Management** (**SIEM**): One of the key advantages of threat intelligence feeds is their seamless integration with SIEM systems. This integration is a game-changer in Recon detection. SIEM systems are designed to collect and analyze vast amounts of data, including logs and security events. By connecting threat intelligence feeds to SIEM, organizations can correlate incoming data with known threat indicators. This correlation empowers early detection of Recon activities. It allows security teams to swiftly recognize and respond to potential threats before they escalate into full-fledged cyberattacks. The synergy between threat intelligence feeds and SIEM systems exemplifies the synergy that's pivotal in modern cybersecurity defense.

- **Open source threat feeds**: Notably, there is a wealth of open source threat intelligence feeds available, such as MISP and OpenDXL. These resources provide organizations with a treasure trove of information on emerging threats and Recon tactics. Open source feeds are a cost-effective way to access valuable threat intelligence. They enable organizations to stay informed about the latest threats and adjust their security measures accordingly.

  Incorporating these feeds into the cybersecurity strategy has been demonstrated to be highly effective. A study by the Cyber Threat Alliance found that organizations leveraging threat intelligence feeds had a significantly higher rate of detecting Recon activities in their early stages. The study emphasized the importance of this approach in strengthening overall cybersecurity posture.

  In conclusion, the utilization of threat intelligence feeds is a proactive and essential component of Recon detection during the cyber kill chain. Their integration with SIEM systems and the availability of open source feeds offer organizations the means to access crucial information about emerging threats, thereby enabling them to detect and respond to Recon activities efficiently.

- **User and Entity Behaviour Analytics (UEBA)**: UEBA is a pivotal element of modern cybersecurity, especially in the context of the Recon stage of the cyber kill chain. UEBA solutions are designed to identify and thwart abnormal behaviors exhibited by users and entities within the network, making them invaluable for recognizing Recon attempts. UEBA leverages advanced methodologies to identify anomalies and potential threats. Key components of UEBA include the following:

  - **Machine learning algorithms**: UEBA leverages sophisticated machine learning algorithms to establish baseline behavior patterns for both users and entities. These algorithms continuously analyze and learn from historical data, enabling them to discern typical behavior. When deviations from these established baselines are detected, it can indicate Recon activities, such as unauthorized access attempts or unusual data access patterns. The application of machine learning ensures that UEBA is adept at recognizing subtle and evolving Recon tactics that may evade traditional rule-based detection.

  - **Contextual analysis**: UEBA solutions stand out for their contextual analysis capabilities. They take into account a wide range of contextual information, including user roles, permissions, and network access patterns. By assessing behavior in the context of the user's role and typical network access, UEBA can make a more accurate determination of whether observed behaviors are indicative of Recon. For example, if a user who typically accesses only a limited set of resources suddenly attempts to access sensitive data or systems beyond their usual scope, UEBA will raise an alert.

  The significance of UEBA in cybersecurity is underscored by the ability to detect insider threats and unknown Recon activities effectively. A study by Gartner found that UEBA solutions provide enhanced visibility into user and entity behavior, resulting in the identification of security incidents that might otherwise go undetected.

UEBA is an essential tool for recognizing Recon activities within the cyber kill chain. Its use of machine learning algorithms and contextual analysis empowers organizations to effectively detect abnormal behaviors among users and entities, ensuring the security of their networks and data.

With a clear understanding of how threat intelligence feeds contribute to identifying potential threats, it's essential to shift our focus towards proactive measures that can effectively prevent Recon activities from compromising our security posture.

Proactive measures to prevent Recon

The Recon phase of a cyberattack is the proverbial first step through the door. To protect against this initial intrusion, organizations must adopt proactive measures that fortify their defenses. In this section, we'll delve into two essential strategies: network segmentation and threat intelligence.

Network segmentation

Network segmentation is a fundamental practice in cybersecurity, serving as a robust defense mechanism, especially during the Recon stage of the cyber kill chain. It involves the strategic division of an organization's network into isolated segments or zones, each equipped with its unique access controls and security measures. The primary objective is to impede lateral movement within the network, significantly increasing the complexity for attackers attempting to navigate the network once they breach its perimeter.

Network segmentation prevents Recon in the following ways:

- **Compartmentalization**: Network segmentation excels in compartmentalizing an organization's infrastructure. In the unfortunate event of a breach, attackers find themselves contained within a single segment. This containment restricts their access to other critical systems and sensitive data, effectively thwarting the spread of their activities. For instance, if an attacker gains access to one segment, they are isolated from the core network, minimizing the damage they can inflict.

- **Access Control**: Each network segment can be fortified with specific access controls and stringent authentication requirements. This means that only authorized personnel can access particular resources within a given segment. This meticulous control reduces the attack surface, ensuring that if Recon activities are initiated within one segment, the attacker's reach remains limited.

- **Visibility**: Network segmentation significantly enhances visibility throughout the network. Suspicious activities within a segment become more conspicuous. The separation of network traffic and resources into distinct segments enables security teams to more effectively detect and respond to Recon attempts. By segregating network elements, it's easier to pinpoint unusual or unauthorized behaviors, enhancing the organization's ability to identify potential threats.

- **Data protection**: A key benefit of network segmentation is its ability to safeguard critical data. Highly sensitive information can be placed within extensively segmented areas, fortified with stringent security measures. This multi-layered protection strategy makes it exceptionally challenging for attackers to access and exfiltrate valuable and confidential information. The isolation of data within secure segments acts as a formidable barrier, impeding Recon efforts aimed at compromising critical assets.

The significance of network segmentation in cybersecurity is underscored by research conducted by the **Cybersecurity and Infrastructure Security Agency (CISA)**. Their study found that organizations implementing network segmentation strategies experienced a significant reduction in the risk of lateral movement by cyber attackers and increased overall network security. The Target breach in 2013 serves as a reminder of the importance of network segmentation. Attackers breached Target's network and, due to inadequate segmentation, were able to move laterally to POS systems, resulting in a massive data breach.

Threat intelligence

Threat intelligence stands as a strategic defense against Recon activities, forming an integral part of an organization's cybersecurity strategy. It encompasses the systematic process of collecting, analyzing, and disseminating information about cyber threats and vulnerabilities. Threat intelligence draws data from diverse sources, including open source feeds, governmental agencies, and industry-specific reports, to provide organizations with a comprehensive understanding of the threat landscape.

Threat intelligence prevents Recon in the following ways:

- **Early warning**: A primary function of threat intelligence is to provide early warnings about emerging threats and trends. This proactive approach equips organizations with the knowledge needed to prepare for known Recon techniques and tactics. By staying ahead of the curve, organizations can thwart potential threats before they escalate into full-scale cyberattacks.

- **IoC**: Monitoring IoCs is a critical aspect of threat intelligence. IoCs, such as suspicious IP addresses and malicious URLs, serve as red flags for Recon activities. By constantly tracking and analyzing these IoCs, organizations can swiftly identify Recon activities and take immediate action to block or mitigate potential threats. This level of vigilance and responsiveness is essential in preventing Recon from evolving into more serious cyberattacks.

- **Attack attribution**: Threat intelligence often includes information about threat actors, their tactics, techniques, and motivations. This knowledge enables organizations to anticipate the types of Recon they might encounter. By understanding the motives and methods of potential adversaries, organizations can adapt their defenses to specifically counter the Recon tactics that are most relevant to their unique circumstances.

- **Customized Defense**: One of the significant advantages of threat intelligence is its ability to enable organizations to customize their defense strategies. It allows organizations to tailor their defenses based on the specific threats that are most pertinent to their industry or region.

Rather than employing a one-size-fits-all approach, organizations can develop highly targeted defenses that are precisely aligned with the Recon activities that pose the most significant risk to them. This level of customization enhances the effectiveness of cybersecurity measures.

The Stuxnet malware, which targeted Iran's nuclear facilities, serves as a prime example of how thorough Recon can empower an exact and sophisticated offensive operation. The attackers leveraged detailed knowledge gained from their Recon efforts to design Stuxnet with unparalleled accuracy, enabling it to exploit specific vulnerabilities within the targeted systems. This case underscores the critical role that comprehensive recon plays in allowing attackers to craft tailored and effective malware, showcasing the offensive potential of such intelligence.

Summary

In this chapter, we delved into the Recon phase of the Cyber Kill Chain, emphasizing its critical significance in the overall process of cyber attacks. Recon, the first stage in a cyber attack, involves gathering as much information as possible about the target, as we discussed in this chapter. Understanding this phase is crucial as it sets the foundation for the attacker's subsequent actions. We learned this in this chapter.

We explored both passive and active Recon methods. Passive Recon involves collecting information without directly interacting with the target, using publicly available data, as well as open source intelligence, as we learned in this chapter. On the other hand, active Recon involves direct interaction with the target's systems to gather more specific details, often raising the risk of detection. We also discussed this in this chapter.

Real-life incidents serve as stark reminders of the devastating effects of successful Recon, driving home the importance of robust defenses. In this chapter, we delved into high-profile breaches during which attackers meticulously gathered information before launching their attacks, shedding light on the intricate planning involved in such operations.

Threat intelligence feeds are pivotal in identifying and mitigating potential threats during the Recon phase, as we learned in this chapter. These feeds help organizations stay ahead of emerging threats by analyzing patterns and IoCs. However, relying solely on threat intelligence is not enough.

We then transitioned to proactive measures to prevent Recon. These measures include implementing strong access controls, regularly updating and patching systems, and employing network segmentation to limit the attack surface. Deceptive technologies, such as honeypots, can also mislead and deter attackers. We covered these in this chapter as well.

By fusing threat intelligence with proactive defenses, organizations can significantly diminish the risk of successful Recon, thereby fortifying their overall security posture. In the Cyber Kill Chain, once reconnaissance is complete, the adversary leverages the intelligence gathered to advance into the *Weaponization* phase. With a deep understanding of the target's vulnerabilities—whether they lie in unpatched systems, misconfigurations, or human elements—the attacker assembles customized tools or exploits. This transition marks the point where passive information gathering gives way to more active intent, as the data collected now serves to craft malware, phishing campaigns, or other means of initial access. At this stage, the adversary's focus shifts from preparation to execution, setting the stage for delivering the payload and establishing a foothold in the target environment. In the next chapter, we will learn about weaponization.

Further readings

Following are some bonus reading materials for you to check out:

- *Symantec report on Stuxnet 0.5, an early version of the Stuxnet worm targeting Iranian nuclear facilities*: https://nsarchive2.gwu.edu/NSAEBB/NSAEBB424/docs/Cyber-088.pdf

- *Article describing how Target's 2013 data breach originated from credentials stolen from an HVAC vendor*: https://krebsonsecurity.com/2014/02/target-hackers-broke-in-via-hvac-company/

- *McAfee Labs blog providing threat research and intelligence from McAfee's threat research team*: https://www.mcafee.com/blogs/other-blogs/mcafee-labs/

- *NIST Special Publication on intrusion detection and prevention systems*: https://nvlpubs.nist.gov/nistpubs/legacy/sp/nistspecialpublication800-94.pdf

- *Paper on implementing full packet capture for network forensics*: https://www.giac.org/paper/gcia/9924/implementing-full-packet-capture/130794

- *Website of the Ponemon Institute, which conducts independent research on data protection and information security*: https://www.ponemon.org

- *Gartner market guide for user and entity behavior analytics (UEBA) solutions*: https://www.business-iq.net/assets/3524-gartner-guide-market-guide-for-user-and-entity-behavior-analytics?open_form=1

- *DoD guide on network segmentation best practices for cloud environments*: https://media.defense.gov/2024/Mar/07/2003407861/-1/-1/0/CSI-CLOUDTOP10-NETWORK-SEGMENTATION.PDF

- *Senate committee report on Target's 2013 data breach and missed opportunities to prevent it*: https://www.commerce.senate.gov/2014/3/rockefeller-staff-report-details-target-s-missed-opportunities-to-stop-massive-data-breach

- *Symantec's technical analysis of the Stuxnet worm targeting industrial control systems*: http://large.stanford.edu/courses/2011/ph241/grayson2/docs/w32_stuxnet_dossier.pdf

Get This Book's PDF Version and Exclusive Extras

UNLOCK NOW

Scan the QR code (or go to packtpub.com/unlock). Search for this book by name, confirm the edition, and then follow the steps on the page.

Note: Keep your invoice handly. Purchase made directly from Packt don't require one.

3

Weaponization

The *Weaponization* phase, the second stage in the cyber kill chain framework, is a critical component of cyberattacks. In this phase, attackers transform identified vulnerabilities and exploits into weaponized malware, setting the stage for the exploitation of target systems.

Weaponization is the critical second phase in the cyber kill chain, a framework that outlines the stages of a cyberattack. In this phase, attackers transition from intelligence gathering to creating their exploitation tools. This stage involves developing malicious payloads designed to exploit vulnerabilities identified during reconnaissance. Attackers craft sophisticated malware—viruses, worms, ransomware—and package these threats within seemingly innocuous formats such as emails, documents, or executable files. The goal is to create a weaponized payload that bypasses detection and maximizes the chances of successful exploitation.

This meticulously planned phase underscores the importance of understanding the target's environment and weaknesses. Attackers employ advanced techniques such as obfuscation and encryption to evade security systems, often using custom-built malware or adapting existing tools to their specific needs. The sophistication of these payloads varies widely, from essential phishing emails to complex zero-day exploits, depending on the attacker's resources and objectives.

In this chapter, we delve into the methodologies and tools used in weaponization, providing crucial insights into how cyber defenders can anticipate and mitigate these evolving threats. This knowledge is a potent weapon in the hands of defenders, empowering them to stay one step ahead of their adversaries and protect their systems effectively.

We'll cover the following key topics:

- Process of weaponization
- Types of malware and payloads
- Case studies – notable weaponization attacks
- Vulnerabilities and exploits used in weaponization
- Strategies to counter weaponization threats

Process of weaponization

The **weaponization phase** in the cyber kill chain is where attackers turn the information they've gathered into powerful cyber weapons. During this phase, they create malicious payloads by exploiting vulnerabilities they found earlier. Attackers often use advanced techniques such as hiding or encrypting their malicious code to avoid detection. The goal is to develop a payload that can silently break through a target's defenses. In this chapter, we will go through the various methods and tools attackers use to create these digital weapons.

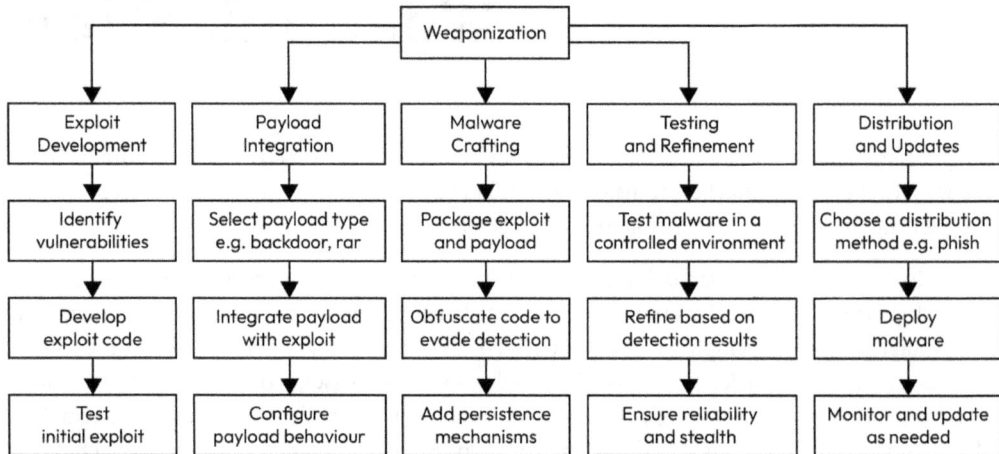

```
                              ┌──────────────┐
                              │ Weaponization │
                              └──────────────┘
```

Exploit Development	Payload Integration	Malware Crafting	Testing and Refinement	Distribution and Updates
Identify vulnerabilities	Select payload type e.g. backdoor, rar	Package exploit and payload	Test malware in a controlled environment	Choose a distribution method e.g. phish
Develop exploit code	Integrate payload with exploit	Obfuscate code to evade detection	Refine based on detection results	Deploy malware
Test initial exploit	Configure payload behaviour	Add persistence mechanisms	Ensure reliability and stealth	Monitor and update as needed

Figure 3.1 – Weaponization process flow

The important steps to carry out the weaponization process are as follows:

1. **Exploit development**:

    - Attackers identify weaknesses in the target system, ranging from software flaws to unpatched vulnerabilities, including zero-day vulnerabilities (unknown to the software vendor).

    - This requires a deep understanding of how software and systems work or access exploits from underground markets and online platforms, where cybercriminals buy and sell various cyber weapons, including exploits, malware, and other tools. The process involves analyzing the vulnerability, creating a proof of concept (a simple demonstration of the exploit), and refining the code to ensure reliability and effectiveness.

2. **Payload integration**:

    - Attackers design a malicious payload, such as a backdoor or data theft tool, and integrate it with the exploit code.

    - The payload is designed to execute seamlessly once the exploit is successful.

    - The exploit and payload are tested and refined to ensure they work as intended without being detected by security systems.

3. **Malware crafting**:

- With the exploit in hand, attackers craft malware, the software that delivers the exploit to the target.

- Malware can take various forms, such as viruses, worms, or spyware, each designed for a specific goal, such as stealing data or controlling a system.

- This process involves setting clear objectives for the malware, designing its structure, writing the main code, and integrating evasion techniques to bypass security measures.

- Testing for **indicators of compromise (IoCs)**: Malware authors rigorously test their creations for any signs or IoCs. These are the digital footprints that security systems use to identify malicious activities. They test their bypasses and evasion techniques in offline and online simulated environments, ensuring the malware can evade detection in real-world scenarios.

4. **Testing and refinement**:

- Malware is extensively tested in isolated environments to ensure it functions correctly across different systems and configurations.

- Attackers refine the malware to improve its stealth and effectiveness, making it harder for security tools to detect.

5. **Distribution and updates**:Malware is packaged and prepared for distribution, often through phishing emails or exploit kits.

Attackers plan to update the malware regularly, ensuring it can adapt to and evade new security measures over time. This dynamic nature of cyber threats underscores the need for continuous vigilance and adaptation in cybersecurity defense. You might encounter some complex or unfamiliar ideas as we continue exploring cyber weaponization and malware. Don't worry—this is normal when diving into such an advanced field. We will explore many of the technical terms, tools, and techniques mentioned here in greater detail in this book. For now, focus on understanding the big picture and how the different elements of weaponization fit together. We'll revisit and expand on these concepts later in the book, helping you build a robust and layered understanding of cybersecurity threats and defenses. By the end of this journey, you'll have a comprehensive knowledge base to draw upon in your professional work.

Next, we'll explore the different types of malware that attackers create, each with unique objectives and challenges. Understanding these will help you better prepare for and defend against the various threats that exist in the cybersecurity landscape.

Common types of malware and payloads

The weaponization phase of the cyber kill chain is a critical stage where attackers make strategic decisions about the types of malware and payloads to employ. These choices are influenced by the attacker's objectives and the vulnerabilities they aim to exploit. In this section, we delve into various common types of malware and payloads that play pivotal roles in this phase:

1. **Trojans**: Trojans are deceptive malware that disguise themselves as legitimate software but contain malicious payloads. Their versatility makes them valuable tools for attackers. Trojans can be used to establish backdoors, steal sensitive information, or initiate further attacks. An example of a Trojan's evolution is the Emotet Trojan, which initially operated as a banking Trojan but later transformed into a versatile delivery mechanism for other malware. It served as a conduit for delivering payloads such as TrickBot and Ryuk ransomware. Another example is the Zeus Trojan, which targets financial information by logging keystrokes and capturing form data, leading to significant financial theft. Emotet, initially a banking Trojan, has evolved into a highly modular malware that distributes other threats, including ransomware. Trojan downloaders, such as Nemucod, download and install additional malicious software on compromised systems. The Rakhni Trojan is particularly versatile, capable of delivering either ransomware or cryptocurrency miners based on the target system's specifications. FakeAV Trojans trick users into believing their system is infected, prompting them to purchase fake antivirus software. These examples underscore the diverse tactics and severe impacts of Trojan malware, highlighting the need for vigilant cybersecurity measures to defend against such threats.

2. **Ransomware**: Ransomware is a type of malware that encrypts a victim's data, rendering it inaccessible until a ransom is paid. Ransomware attacks have gained notoriety for causing significant disruptions and financial losses to organizations and individuals. These attacks often exploit vulnerabilities in the victim's system to gain initial access. Famous examples include WannaCry, which in May 2017 exploited a vulnerability in Windows systems to infect over 200,000 computers across 150 countries, severely impacting services such as the UK's National Health Service. Another infamous variant is Petya/NotPetya, which initially targeted the master boot record to encrypt the file system table. Still, later versions, such as NotPetya, caused widespread disruption by acting as a wiper rather than ransomware, notably affecting many companies in June 2017. Ryuk ransomware has recently targeted large enterprises, encrypting critical files and demanding high ransoms, often delivered through phishing emails or other malware such as TrickBot. These examples highlight ransomware attacks' evolving nature and significant impact on global businesses and infrastructure.

3. **Rootkits**: Rootkits are designed to hide the presence of malware on compromised systems. They manipulate the **operating system (OS)** to evade detection and maintain persistence. The Rustock rootkit, which infected Windows-based systems, was notorious for its ability to conceal itself by tampering with the Windows kernel. Rootkits are particularly effective in establishing and maintaining covert access to compromised systems. Notable examples include the Sony BMG rootkit, which, in 2005, was covertly installed on millions of computers through Sony's music CDs, leading to significant privacy and security concerns and legal actions. Stuxnet, which used a

sophisticated rootkit, targeted Iran's nuclear facilities and manipulated industrial control systems while concealing its presence. Another example is TDSS/TDL-4, a highly advanced rootkit that infected millions of computers worldwide, employing sophisticated techniques to hide from detection and remove antivirus software. TDSS/TDL-4 stands for "Tidserv" or "Alureon" Rootkit, with TDL-4 referring to the 4th generation of the TDL (Teldra or Alureon) rootkit family. This rootkit is notorious for its stealth and ability to bypass security measures, making it a persistent threat on infected systems. These rootkits illustrate the significant threat they pose to security and privacy, often requiring specialized tools and methods for detection and removal.

4. **Remote access Trojans (RATs)**: RATs provide attackers with complete control over compromised systems, enabling various activities such as data theft, surveillance, or further exploitation. The Poison Ivy RAT, used in various cyber espionage campaigns, allowed attackers to remotely control infected systems and exfiltrate sensitive data. RATs are favored by attackers seeking persistent and surreptitious access. Prominent examples include **DarkComet**, which has been widely used since its creation in 2008 and offers features such as keylogging, screen capturing, and remote desktop access, making it a favorite among cybercriminals. Another example is **Gh0st RAT**, which originated in China and has been used in various cyber espionage campaigns, notably targeting government and corporate entities to steal sensitive information. These RATs illustrate the significant risk they pose by providing attackers with extensive control over compromised systems, often leading to severe data breaches and prolonged unauthorized access.

5. **Exploit kits**: Exploit kits are bundles of pre-packaged exploits that simplify the process of targeting known software vulnerabilities. Attackers often distribute them through malicious websites, taking advantage of visitors' unpatched software. Notable examples include the **Angler exploit kit**, which was highly active until its shutdown in 2016. It is known for exploiting vulnerabilities in Adobe Flash, Java, and Silverlight to deliver ransomware and other malware. The **Nuclear exploit kit** targeted vulnerabilities in web browsers and plugins to install banking Trojans and ransomware before its operation ceased in 2016. Another prominent example is the **RIG exploit kit**, which remains active and continues to evolve, leveraging various exploits to distribute malware such as CryptXXX ransomware. These exploit kits illustrate the cyberattack's automated and scalable nature, enabling attackers to target numerous victims by exploiting common software vulnerabilities.

6. **Fileless malware**: Fileless malware operates in system memory without leaving traces on the victim's system, making it particularly challenging to detect and remove. Famous examples include PowerShell-based attacks, where attackers use legitimate PowerShell scripts to execute malicious activities directly in memory. One such attack was observed with APT29, a group linked to Russian intelligence, using PowerShell to conduct espionage. Another example is Duqu 2.0, which exploited zero-day vulnerabilities to operate entirely in memory, targeting high-profile entities such as Kaspersky Lab and the Iranian nuclear talks. Poweliks is a well-known fileless malware that resides in the Windows registry and uses JavaScript and PowerShell to execute its payload. These examples highlight the stealth and sophistication of fileless malware, posing significant challenges for traditional antivirus solutions and necessitating advanced detection techniques. Fileless malware leverages legitimate system processes and scripting languages to carry out malicious activities, making it a potent tool for attackers.

7. **Adware**: Adware, a seemingly less harmful software, is designed to automatically display or download advertisements on a user's computer, often without their consent. However, it's crucial to understand that adware can degrade system performance and, more critically, serve as a gateway for more dangerous malware. The Fireball adware, for instance, not only invaded millions of systems worldwide but also had the potential to install additional malicious software, manipulate web traffic, and even execute arbitrary code on the victim's machine. This highlights the need for caution and awareness when dealing with seemingly harmless adware.

8. **Spyware**: Spyware, a stealthy and covert threat, gathers information from an infected system without the user's knowledge. This information can include sensitive data such as login credentials, financial information, and personal communications. The Pegasus spyware, developed by the NSO Group, is a prime example. It can be installed on a victim's device through a simple phishing attack or exploiting zero-day vulnerabilities, granting the attacker complete access to the device's data. This underscores the need for vigilance and proactive measures to protect against such stealthy threats.

9. **Worms**: Worms, rapidly spreading malware, self-replicate and spread across networks, often exploiting vulnerabilities to propagate without needing human intervention. The WannaCry ransomware worm is a notorious example, leveraging the EternalBlue exploit to rapidly infect thousands of systems globally, causing widespread disruption and financial loss. The rapid spread of worms such as WannaCry should serve as a stark reminder of the need to be prepared and vigilant against such highly destructive malware.

10. **Botnets**: Botnets are networks of compromised computers controlled by an attacker. They are often used to launch large-scale attacks such as **distributed denial of service (DDoS)** attacks, spam campaigns, or further malware distribution. The Mirai botnet, which targeted IoT devices, is a prime example. It caused massive DDoS attacks, temporarily taking down major websites such as X (formerly Twitter) and Reddit. Botnets exemplify how attackers can leverage many compromised systems to amplify attacks, making them difficult to mitigate and highly destructive.

11. **Rogue security software**: Rogue security software, also known as **scareware**, pretends to be legitimate antivirus or security software but is designed to trick users into paying for fake or unnecessary services. Once installed, it may generate false alerts and demand payment to remove non-existent threats. An example is the Mac Defender scareware, which targeted Mac users by redirecting them to fake security alerts and prompting them to install malicious software disguised as legitimate antivirus programs.

The choice of malware and payloads during the weaponization phase is a critical strategic decision for attackers. They carefully select the type of malware that aligns with their goals and the vulnerabilities they intend to exploit. Understanding these common types of malware and payloads is essential for organizations and security professionals to enhance their defenses and respond effectively to cyber threats. Next, let's look into some important case studies of weaponization attacks.

Case studies – notable weaponization attacks

To gain a deeper understanding of the *Weaponization* phase in cyberattacks, it's important to examine specific incidents where weaponization played a pivotal role. In this section, we will explore several case studies that highlight the significance of weaponization in cyberattacks.

Case study 1 – NotPetya – disguised as ransomware, inflicted destruction

This case study delves into the origins, mechanics, and devastating consequences of malware that initially posed as ransomware but had far more sinister objectives.

The NotPetya attack, originating in Ukraine, initially bore the appearance of ransomware. It arrived in the form of a ransomware strain, demanding a Bitcoin payment in exchange for a decryption key. This tactic was not unusual; ransomware attacks had been on the rise, with attackers typically seeking financial gain.

However, what set NotPetya apart was its true intent. It was not designed to facilitate financial extortion but to wreak havoc and destruction on an unprecedented scale. In essence, it was a wolf in sheep's clothing, designed to deceive both its victims and the cybersecurity community.

NotPetya was a carefully engineered piece of weaponized malware. It leveraged a well-known ransomware strain, Petya, as a disguise. The attack was propagated through a malicious update to a popular Ukrainian tax accounting software, MeDoc. Once installed on a system, NotPetya encrypted the user's data, making it inaccessible, and displayed a ransom note, demanding payment in exchange for a decryption key.

However, the payment was a ruse. Unlike typical ransomware attacks, where victims could regain access to their data after paying the ransom, NotPetya's encryption was irreversible. Paying the ransom would not result in data recovery. This malicious deception marked a significant departure from the typical modus operandi of ransomware attacks, as it was never about the money.

The malware's destructive power was further amplified by its ability to propagate within networks. It leveraged an exploit called EternalBlue, the same exploit used by the WannaCry ransomware, which allowed it to spread rapidly through unpatched or vulnerable systems. This enabled NotPetya to move laterally within organizations, causing widespread infection.

The NotPetya attack targeted businesses, particularly in Ukraine. However, its impact extended far beyond its initial targets. NotPetya was not content with merely encrypting data; it aimed to cripple organizations and disrupt critical infrastructure. As a result, it caused significant financial losses and operational disruptions that rippled across the globe.

Among the organizations most severely affected were Ukraine's critical infrastructure, including its power grid, airports, banks, and government systems. It resulted in widespread chaos and panic, revealing the vulnerability of critical infrastructure to such attacks.

Global corporations with subsidiaries or partners in Ukraine also suffered tremendous losses. These multinational companies experienced severe operational disruptions and financial setbacks, highlighting the interconnected nature of the global business landscape.

The NotPetya attack drew widespread attention and condemnation from the international community. The Ukrainian government, in particular, bore the brunt of the attack's initial impact and worked tirelessly to respond and recover.

In the aftermath of the attack, cybersecurity researchers and experts across the world joined forces to investigate and attribute the attack. While the perpetrators initially posed as a ransomware gang, it soon became apparent that the attack had a more significant, state-sponsored origin.

NotPetya was attributed to the Russian military, specifically the Russian **GRU** (in Russian: **Glavnoe Razvedyvatelnoe Upravlenie**) intelligence agency. The attack was seen as a deliberate act of aggression against Ukraine, as well as a display of cyber capabilities and an attempt to destabilize the region.

The NotPetya attack served as a wake-up call for governments, organizations, and the cybersecurity community. It shattered the notion that ransomware attacks were solely about financial gain. NotPetya demonstrated the potential for cyberattacks to cause irreversible damage and disruption on a scale that could rival traditional acts of war.

One of the primary lessons from the NotPetya attack was the critical importance of patch management and cybersecurity hygiene. The malware's use of the EternalBlue exploit highlighted the need for organizations to keep their systems up to date and promptly apply security patches. Failure to do so can leave systems vulnerable to exploitation.

The attack also underscored the interconnected nature of the global business landscape. A cyberattack originating in one region could have far-reaching consequences for organizations and critical infrastructure worldwide. The NotPetya attack exposed the need for international collaboration and information sharing to mitigate and respond to such threats effectively.

NotPetya, the deceptive ransomware that concealed its true intent, served as a stark reminder of the evolving nature of cyber threats. Its devastating impact on Ukraine's critical infrastructure and global corporations highlighted the need for robust cybersecurity measures, international cooperation, and the imperative of maintaining systems' security.

The attack forever changed the perception of ransomware, revealing that cyber threats could extend beyond financial extortion and possess the capacity to inflict widespread chaos and destruction. The NotPetya case study stands as a testament to the ever-evolving threat landscape of the digital age.

Case study 2 – Emotet – a versatile malware delivery platform

Emotet stands as a case study in versatility, persistence, and the profound implications it holds for the security of individuals and organizations worldwide.

Emerging on the cyber threat scene in 2014, Emotet initially presented itself as a banking Trojan. Its primary objective was to steal financial information and credentials, a malicious intent aligned with traditional cybercrime. However, what sets Emotet apart is its evolution into a versatile malware delivery platform. Rather than conducting attacks directly, Emotet serves as a delivery mechanism for a wide range of malicious payloads.

Emotet's primary mode of propagation is through spam emails. Cybercriminals employ deceptive subject lines and often employ well-crafted attachments or links to dupe unsuspecting recipients into opening these files. Once a system is compromised, Emotet takes control, transforming it into a **command-and-control** (**C2**) server. This establishes a direct line for attackers to remotely control the infected system and distribute further malware.

Emotet's distinguishing feature lies in its modular design. This architecture allows attackers to adapt and update their payloads with ease. Essentially, Emotet functions as a versatile malware **loader**, making it an ideal platform for deploying a variety of malicious payloads. These payloads can range from banking Trojans designed for financial fraud to ransomware, the notorious form of malware that encrypts victims' data and demands a ransom for its release.

What makes Emotet particularly challenging for defenders is its polymorphic nature. **Polymorphism** is a technique where the malware's code changes with each infection, rendering traditional antivirus solutions less effective. As a result, Emotet is continually morphing, making it challenging to detect and mitigate.

One of the most significant and notorious roles Emotet has played is as a delivery platform for ransomware. Ransomware attacks have had a profound impact on both businesses and individuals. These attacks involve the encryption of data on the victim's system, followed by a ransom demand for its decryption.

Emotet has facilitated the distribution of various ransomware strains, including Ryuk and TrickBot. Ryuk, in particular, has become synonymous with devastating and high-stakes ransomware attacks. These attacks have resulted in significant financial losses, data breaches, and operational disruptions across many sectors, from healthcare to critical infrastructure.

Emotet's operations spanned the globe, affecting organizations and individuals in numerous countries. It was a persistent and international threat, highlighting the interconnected nature of cybersecurity challenges. The malware's operators expertly evaded detection, using a combination of techniques to maintain their infrastructure and keep their malicious campaigns running.

In a landmark development, international law enforcement agencies coordinated efforts to dismantle the Emotet infrastructure in early 2021. This effort involved seizing control of the servers used by the malware, disrupting its operations. The takedown marked a significant achievement in the fight against cyber threats. However, it also raises questions about the future of Emotet and whether it will be permanently defanged or if alternative delivery mechanisms will emerge.

Emotet's legacy underscores the need for ongoing vigilance in the face of cyber threats. Its adaptability and persistence serve as a stark reminder of the ever-evolving nature of cybersecurity challenges. The malware's role as a ransomware delivery platform is particularly noteworthy, as it has contributed to significant data loss, financial losses, and operational disruptions for countless entities.

The Emotet case study highlights the importance of adopting robust security measures. Cybercriminals will continue to leverage malware delivery platforms, and organizations must invest in comprehensive cybersecurity solutions to defend against these threats. Moreover, the international collaboration that led to the takedown of Emotet showcases the need for coordinated efforts to combat cybercrime on a global scale.

The Emotet story is not just one of a malware platform but a testament to the ongoing challenges in cybersecurity. It emphasizes that cybersecurity is a shared responsibility, and the fight against cyber threats requires a global, collaborative effort.

Emotet, once a banking Trojan, morphed into a versatile malware delivery platform that facilitated a range of cyber threats, most notably ransomware attacks. Its adaptability, modularity, and ability to avoid detection make it a formidable adversary in the world of cybersecurity. The international takedown of Emotet's infrastructure marked a significant achievement but also underscored the need for ongoing vigilance in the face of evolving cyber threats. Emotet's legacy serves as a compelling case study in the ever-evolving landscape of cybersecurity.

Vulnerabilities and exploits used in weaponization

Attackers rely on specific vulnerabilities and exploits to achieve their objectives. Understanding these vulnerabilities and exploits is crucial for defenders seeking to fortify their systems against cyber threats. In this section, we will explore common vulnerabilities and exploits that attackers leverage during this stage.

Common vulnerabilities

As we transition from discussing various forms of malware, it becomes crucial to understand the underlying vulnerabilities and exploits leveraged in the weaponization phase of cyberattacks. These vulnerabilities are the Achilles' heel that attackers exploit, turning seemingly benign software into powerful weapons capable of wreaking havoc on systems and networks. By examining the specific exploits and techniques used in notable cases, we can gain deeper insights into how attackers craft their malicious payloads and the critical importance of robust cybersecurity measures to defend against such sophisticated threats.

- **Software vulnerabilities**: Attackers frequently target software vulnerabilities as an entry point into systems. These vulnerabilities encompass a wide spectrum, ranging from unpatched security flaws to design weaknesses that enable unauthorized access. The timely application of software patches and updates is critical in mitigating this type of risk. Vulnerabilities in software applications and OSs provide attackers with a foothold, and once they breach these weaknesses, they can execute a variety of malicious actions. Such actions may include stealing sensitive data, launching denial-of-service attacks, or gaining unauthorized control over systems. The mitigation strategy here is clear: organizations must establish rigorous patch management processes to address software vulnerabilities promptly.

- **Network misconfigurations**: Network misconfigurations, primarily within network devices such as routers and firewalls, represent a common source of vulnerability. Errors in the configuration of these devices can inadvertently expose vulnerabilities that attackers can exploit. Misconfigured access control lists or firewall rules may unintentionally permit unauthorized access, leaving networks susceptible to breaches. Addressing network misconfigurations involves meticulous device configuration reviews, continuous monitoring, and rigorous access control measures. Identifying and rectifying these errors is essential in fortifying network security.

- **Social engineering**: While not a traditional software vulnerability, social engineering tactics exploit human weaknesses to gain unauthorized access to systems. Social engineering tactics, such as phishing emails, rely on psychological manipulation to deceive users into revealing sensitive information or executing malicious code. These tactics often bypass technical security measures, making them a formidable threat. Mitigating social engineering vulnerabilities necessitates comprehensive user training and awareness programs. It is essential to educate users about the dangers of social engineering and equip them with the skills to identify and respond to such attacks. Implementing robust email filtering and authentication mechanisms can also be effective in reducing the success of phishing campaigns.

Understanding common vulnerabilities is the first step in fortifying our defenses against cyber threats. With this foundation, we can now delve into the specific exploits that attackers use to leverage these weaknesses, transforming potential security gaps into active threats. By examining how these exploits are crafted and deployed, we can better appreciate the intricate methods used in cyberattacks and enhance our strategies to counter them effectively.

Common exploits

It is essential for a cybersecurity professional to understand the specific exploits that attackers use to take advantage of the vulnerabilities. These exploits serve as the tools and techniques that transform potential vulnerabilities into active threats, allowing attackers to breach systems, execute malicious code, and achieve their objectives. By focusing on common exploits, we not only gain insight into the methods employed by cybercriminals but also significantly enhance our ability to detect, mitigate, and prevent such attacks:

- **Zero-day exploits**: Zero-day exploits are a formidable category of exploits, as they target vulnerabilities that are unknown to the software vendor or have not yet been patched. These vulnerabilities are called *zero days* because there are zero days of protection in place. Attackers, thus, have the upper hand as they can exploit these weaknesses before any countermeasures are available. For organizations, this underscores the necessity of proactive security measures and rapid response capabilities. Zero-day attacks have been responsible for some of the most significant cybersecurity breaches in recent history. One prominent example is the 2010 Stuxnet worm, which exploited multiple zero-day vulnerabilities in Windows systems and Siemens industrial control software to sabotage Iran's nuclear program. In 2014, the Heartbleed bug, a zero-day vulnerability in OpenSSL, exposed sensitive data across millions of websites. The 2017 EternalBlue exploit, leaked from the **National Security Agency** (**NSA**), was used in the WannaCry and NotPetya attacks before Microsoft could patch the vulnerability. More recently, in 2021, the zero-day vulnerabilities in Microsoft Exchange Server led to widespread compromises of email systems worldwide. The SolarWinds supply chain attack, discovered in late 2020, leveraged a zero-day vulnerability to infiltrate numerous government agencies and private companies. These examples demonstrate the severe impact of zero-day attacks across various sectors and underscore the critical need for robust cybersecurity measures and rapid response capabilities. A study by Symantec on zero-day vulnerabilities highlights their significance, emphasizing the need for comprehensive patch management and intrusion detection strategies.

- **Buffer overflow**: Buffer overflow exploits programming errors that allow attackers to overwrite a program's memory. This manipulation could lead to remote code execution, making buffer overflows a prevalent method for attackers seeking unauthorized access to a system. Mitigating buffer overflow vulnerabilities requires robust code review processes and implementing security mechanisms such as **address space layout randomization** (**ASLR**) and **Data Execution Prevention** (**DEP**). Additionally, adopting memory-safe programming languages, such as Rust, Go, and Java, provides a proactive defense by preventing such memory manipulation errors at the code level, representing a newer recommendation beyond OS-level measures. The Morris worm, a classic example, exploited a buffer overflow vulnerability in the Unix finger daemon 1988. The Heartbleed vulnerability, another prominent case, allowed attackers to read sensitive information from memory. More recently, the EternalBlue exploit, used in the WannaCry ransomware attack, demonstrated how a buffer overflow vulnerability in Microsoft's **Server Message Block** (**SMB**) protocol could be leveraged to execute arbitrary code on target machines. These examples underscore the critical importance of secure coding practices, memory-safe languages, and rigorous testing, as they are undoubtedly key to preventing such devastating exploits.

- **SQL injection**: SQL injection exploits occur when attackers manipulate input fields in web applications to execute arbitrary SQL queries. This manipulation can lead to unauthorized access to databases, enabling attackers to extract, modify, or delete sensitive data. Notably, SQL injection is a well-documented attack vector, and organizations should prioritize secure coding practices, input validation, and parameterized queries to prevent such exploits. The **Open Web Application Security Project** (**OWASP**) provides valuable guidance on mitigating SQL injection vulnerabilities. One prominent example is the 2008 Heartland Payment Systems breach, where attackers used SQL injection to install malware on the company's network, compromising over 100 million credit card numbers. Another significant case involved Yahoo! in 2012, where an SQL injection vulnerability in the Yahoo! Voices website allowed attackers to extract 450,000 unencrypted usernames and passwords, exposing them online. Similarly, the 2014 attack on eBay exploited SQL injection vulnerabilities to gain access to a database containing the personal information of 145 million users. These incidents underscore the devastating impact of SQL injection attacks and highlight the importance of securing web applications against such vulnerabilities through input validation, parameterized queries, and regular security testing.

- **Cross-site scripting (XSS)**: XSS exploits involve injecting malicious scripts into web pages viewed by other users. These scripts can compromise user sessions, leading to the theft of session cookies and other sensitive data. As such, XSS is a prevalent method for attackers looking to compromise web applications. Secure development practices, input validation, and output encoding are key strategies for mitigating XSS vulnerabilities. Additionally, organizations can benefit from the guidance provided by the OWASP XSS *Prevention Cheat Sheet*. One well-known example is the Samy worm from 2005, which exploited a vulnerability on MySpace. The worm added Samy Kamkar's name to over a million profiles by injecting malicious JavaScript code into his profile page, which spread to anyone who viewed it. Another significant incident occurred with eBay in 2014, where attackers used XSS vulnerabilities to redirect users to phishing sites designed to steal login credentials. Additionally, Twitter faced an XSS attack in 2010, known as the onMouseOver exploit, where attackers injected malicious code that executed when users hovered over a tweet. These examples highlight the potential impact of XSS vulnerabilities and the necessity for robust input validation and output encoding to mitigate such risks.

- **Phishing exploits**: Phishing attacks are distinct in that they don't target software vulnerabilities but rather exploit human psychology. Attackers craft convincing emails or websites that appear legitimate, deceiving users into revealing sensitive information or executing malicious code. The Human Factors and Ergonomics Society has conducted research on user behavior in phishing attacks, emphasizing the importance of user education and awareness to combat these social engineering exploits. One infamous example is the 2016 **Democratic National Committee (DNC)** breach, where attackers used spear-phishing emails to trick DNC officials into revealing their credentials, leading to a significant data breach and the leak of confidential emails. Another notable case is the Sony Pictures hack in 2014, where phishing emails were used to access the company's network, resulting in the theft and release of vast amounts of sensitive data, including unreleased films and personal employee information. Additionally, the 2013 Target data breach involved attackers using phishing emails to gain access to the credentials of a third-party **Heating, Ventilation, and Air Conditioning (HVAC)** contractor, which allowed them to infiltrate Target's network and steal credit card information from millions of customers. These incidents underscore the severe impact of phishing attacks and the critical need for robust email security and user awareness training.

A deep understanding of common exploits is important to effective cybersecurity, particularly during the reconnaissance stage of the cyber kill chain. This knowledge enables organizations to implement proactive measures, apply necessary patches, and educate users on recognizing and countering social engineering attacks, thus bolstering their overall security posture.

Strategies to counter weaponization threats

Cybersecurity is marked by a constant arms race between defenders and adversaries, with the weaponization phase in the cyber kill chain representing a critical battleground. This stage is where cyberattackers weaponize identified vulnerabilities to launch their attacks. Counteracting weaponization threats is paramount for organizations aiming to safeguard their digital assets and sensitive information.

Organizations must adopt a comprehensive and multi-layered approach encompassing several critical strategies to counter weaponization threats effectively. Vulnerability management is foundational, involving regular patching and scanning to identify and remediate security weaknesses before attackers can exploit them. Application allowlisting ensures that only authorized software runs on systems, significantly reducing the risk of executing malicious code. Endpoint security solutions, such as advanced anti-malware and **Endpoint Detection and Response (EDR)** systems, provide robust defenses against sophisticated threats targeting individual devices. Email security measures, including phishing protection and attachment/link analysis, help prevent malicious content from reaching users. Finally, fostering a culture of security awareness through regular training and incident response drills empowers employees to recognize and respond to potential threats effectively. By integrating these strategies, organizations can create a resilient security posture capable of mitigating the risks posed by weaponization threats.

Vulnerability management

Effective vulnerability management is essential for safeguarding organizational assets against exploitation in the dynamic landscape of cybersecurity. This process involves identifying, evaluating, treating, and mitigating vulnerabilities in software and hardware to prevent attackers from exploiting these weaknesses. Regularly updating and patching systems, conducting thorough vulnerability assessments, and implementing automated scanning tools are pivotal steps in this strategy. By proactively managing vulnerabilities, organizations can significantly reduce their attack surface and enhance their resilience against potential cyber threats. This systematic approach fortifies defenses and aligns with best practices recommended by industry standards and regulatory frameworks, such as those from the **National Institute of Standards and Technology (NIST)** and **International Organization for Standardization (ISO)**.

Importance of timely patching

Timely patching stands as a cornerstone of cybersecurity. Organizations must prioritize the swift application of security patches and updates for all software and systems in use. Vulnerabilities are often exploited through known security flaws, emphasizing the critical role that patch management plays as a fundamental defense strategy. Cyber adversaries are well versed in exploiting these known vulnerabilities, leaving organizations that delay patching at a heightened risk of exploitation. The *Equifax Data Breach Report* attributed this to the failure to apply a critical patch in a timely manner, illustrating the dire consequences of inadequate patch management.

Automated patching solutions

The landscape of modern computing is characterized by its complexity, with a myriad of software and systems requiring constant attention. In this environment, the use of **automated patching solutions** emerges as a critical strategy. Automated solutions streamline the entire process of identifying, testing, and deploying patches across an organization's infrastructure. By automating these tasks, organizations not only save valuable time and resources but also significantly reduce the window of vulnerability. A study by the **National Vulnerability Database (NVD)** revealed that organizations leveraging automated patch management systems had, on average, a 40% faster patch deployment rate than those relying on manual methods.

Patch validation and testing

While the urgency of patch deployment cannot be overstated, the process must be tempered with caution. It is crucial to validate and rigorously test patches in a controlled environment before deployment to ensure that they do not inadvertently disrupt critical systems or introduce new issues. Rushed deployments can inadvertently lead to operational disruption. The aftermath of the WannaCry ransomware attack in 2017 highlighted this, where some organizations hastily deployed a critical patch, only to discover it caused system instability. To prevent such scenarios, organizations must institute a robust patch-testing methodology, encompassing various scenarios and regression testing, ensuring patches are reliable and do not inadvertently introduce new vulnerabilities.

Automated patching solutions, while saving time and resources, also substantially reduce the window of vulnerability. However, this expeditious approach must be balanced with the rigor of patch validation and testing in a controlled environment to ensure the stability and security of the organization's systems.

Application allowlisting

Application allowlisting is a powerful strategy for enhancing an organization's cybersecurity posture by controlling the software allowed to execute on its systems. It forms a critical defense mechanism against malicious software and unauthorized applications, offering a proactive approach to security.

Trusted applications should be defined as:

- **Application allowlisting**: Creating a well-defined list of trusted applications authorized to run on an organization's systems. This approach contrasts with delisting, as only pre-approved software can execute. By preventing rogue or unauthorized software from infiltrating the system, allowlisting reduces the attack surface.

- **Protection against zero-day exploits**: Application whitelisting offers robust protection against threats, including zero-day exploits, which target vulnerabilities unknown to the software vendor. It allows only known, trusted, and verified applications to operate.

- **Implementation**: Organizations should conduct rigorous application assessment, validation, and approval procedures. This involves evaluating software for legitimacy, security, and necessity within the organization, then adding approved applications to the allowlist.

The default-deny approach is used in the following scenarios:

- **High-level security**: This model categorically prohibits all software execution unless explicitly approved, offering the utmost level of protection.

- **Comprehensive approval process**: Every new application or update must undergo thorough scrutiny to ensure compliance with security standards. Approved software is then added to the allowlist.

- **Industry application**: Commonly employed in sectors where the security of sensitive data is paramount, such as defense, finance, and government organizations.

Continuous monitoring is good for the following reasons:

- **Adapting to changes**: The dynamic nature of software environments requires regular review and updates of application whitelists to accommodate legitimate changes while maintaining security.

- **Automated tools and processes**: Facilitate efficient management of whitelists by recognizing and validating software updates, new versions, and additional applications.

- **Balancing security and adaptability**: Continuous monitoring ensures that the organization remains secure and adaptable to evolving cybersecurity threats.

Application allowlisting represents a proactive and formidable defense against malicious software and unauthorized applications. By meticulously defining trusted applications, implementing the default-deny approach when necessary, and continuously monitoring and adapting allowlists, organizations can significantly enhance their cybersecurity posture. This approach not only minimizes the attack surface but also empowers organizations to combat the most sophisticated cyber threats effectively.

Network security measures

Network security measures play a pivotal role in countering weaponization threats by providing multiple layers of defense that detect, prevent, and respond to malicious activities. One essential measure is the implementation of **intrusion detection and prevention systems (IDPSs)**. These systems monitor network traffic for suspicious patterns that may indicate an attack. By analyzing network packets and comparing them against known threat signatures and anomaly-based detection models, IDPS can identify and block malicious activities before they penetrate deeper into the network. This proactive detection is crucial for stopping weaponized payloads and preventing the initial delivery or propagation of malware such as ransomware, Trojans, and other malicious software.

Key techniques include the following:

- **IDPSs**: Monitor and analyze network traffic to detect and block malicious activities using signature and anomaly-based detection models.

- **Network segmentation**: Divide the network into smaller, isolated segments to limit attackers' lateral movement and contain potential threats.

- **Firewalls**: Control incoming and outgoing network traffic based on predefined security rules to block unauthorized access.

- **Virtual private networks (VPNs)**: Encrypt data transmissions to protect against eavesdropping and secure internal communications.

- **Access controls**: Implement stringent access controls to sensitive data and critical systems to prevent unauthorized access and minimize the impact of potential breaches.

Implementing these network security strategies ensures a robust defense against the sophisticated tactics used in weaponization, thereby safeguarding organizational assets and maintaining the network's integrity.

Endpoint security

Endpoint security measures are crucial in countering weaponization threats, as endpoints are often the primary targets for delivering and executing malicious payloads. Advanced anti-malware solutions provide robust protection by detecting and blocking malware before it can infect the system. These solutions leverage signature-based detection to identify known threats and behavioral analysis to spot suspicious activities indicative of new or unknown malware. Additionally, real-time scanning and automatic updates ensure that endpoints are continuously protected against emerging threats. Endpoint security suites often include firewalls, **intrusion detection systems (IDSs)**, and application control, further fortifying the device against potential attacks. For example, CrowdStrike, SentinelOne, and McAfee offer comprehensive endpoint security solutions that provide layered defenses, enhancing the overall security posture.

Another critical aspect of endpoint security is the implementation of EDR solutions. EDR tools provide continuous monitoring and analysis of endpoint activities to swiftly detect and respond to potential threats. These tools are designed to identify **advanced persistent threats (APTs)** and other sophisticated attacks that may evade traditional security measures. EDR solutions enable security teams to investigate incidents thoroughly, trace the attack's origin, and understand its impact on the system. By providing detailed forensics and real-time alerts, EDR solutions help organizations mitigate threats before they cause significant damage. The comprehensive protection offered by the integration of EDR with other security measures, such as user behavior analytics and threat intelligence, further enhances the ability to counter weaponization threats effectively. Companies such as CrowdStrike and Carbon Black are known for their advanced EDR capabilities, offering comprehensive protection against modern cyber threats.

Key techniques include the following:

- **Advanced anti-malware solutions**: Utilize signature-based detection and behavioral analysis to identify and block known and unknown malware.

- **Real-time scanning and automatic updates**: Keep anti-malware definitions and software up to date to ensure continuous protection against emerging threats.

- **EDR solutions**: Provide continuous monitoring, real-time alerts, and detailed forensics to detect and respond to sophisticated threats.

- **Firewalls and IDSs**: Protect endpoints by controlling network traffic and detecting suspicious activities.

- **Application control**: Limit the execution of unauthorized software to prevent malicious applications from running on endpoints.

- **User behavior analytics**: Monitor user activities to detect anomalies that may indicate a security threat.

- **Threat intelligence integration**: Use threat intelligence feeds to stay informed about the latest threats and adjust security measures accordingly.

Implementing these endpoint security strategies ensures robust protection against the sophisticated tactics used in weaponization, safeguards organizational assets, and maintains endpoint integrity.

Automatic computer locking policies

One of the simplest yet most effective measures to enhance endpoint security is automatically locking computers when idle for a specific period. This policy ensures that if a user steps away from their workstation, the system will automatically lock itself, requiring re-authentication to gain access again. This reduces the risk of an unauthorized person accessing the computer and potentially introducing malware through a USB or other external devices. Organizations should do the following:

- **Set idle time limits**: Implement a policy where systems automatically lock after a brief period of inactivity, such as 5-10 minutes.

- **Require strong authentication**: To add an additional layer of security, ensure that unlocking the computer requires strong authentication methods, such as **multi-factor authentication** (**MFA**).

Hardware allow listing

Implementing a hardware allow list can be a critical control to further protect against threats, such as EternalBlue, that might be introduced via USB devices. A hardware allow list restricts the types of external devices that can connect to a system, only permitting those that are pre-approved and recognized by the organization's security policies. This helps to prevent unauthorized USB devices, which could be used to introduce malware, from being connected to the network:

- **Device control software**: Use software that enforces hardware listing, ensuring only trusted devices can connect to company systems.

- **Regular updates**: Continuously update the allow list to include new authorized devices while promptly removing devices that are no longer in use or are deemed insecure.

Combining these controls into the organization's security strategy can significantly reduce the risk of unauthorized access and malware via physical media, thereby improving the overall security posture.

Email security

Email security measures are essential in countering weaponization threats, as emails are a primary vector for delivering malicious payloads. Phishing protection tools, such as advanced email filters and anti-phishing software, are crucial in identifying and blocking malicious emails before they reach the user's inbox. These tools analyze email content for suspicious links, attachments, and patterns that resemble phishing attempts. By employing machine learning algorithms and threat intelligence feeds, these systems can detect known threats and adapt to emerging ones. Additionally, sandboxing techniques allow email security solutions to open and analyze attachments and links in a secure, isolated environment, ensuring that any malicious behavior is detected and contained before it can affect the user's system. This proactive approach significantly reduces the risk of users inadvertently executing weaponized payloads delivered via email.

Furthermore, email authentication protocols such as **Sender Policy Framework (SPF)**, **DomainKeys Identified Mail (DKIM)**, and **Domain-Based Message Authentication, Reporting, and Conformance (DMARC)** help in verifying the legitimacy of email senders. These protocols ensure that emails are not spoofed and that the domain from which an email is sent is authorized to send emails on behalf of that domain. User awareness and training are also critical components of email security. Regular training sessions can educate users about recognizing phishing emails, the dangers of clicking on unknown links or attachments, and the importance of reporting suspicious emails. Combined, these measures form a robust defense against email-borne weaponization threats, significantly mitigating the risk of successful cyberattacks initiated through email channels.

Key techniques include the following:

- **Phishing protection tools**: Advanced email filters and anti-phishing software that detect and block malicious emails. Some of the popular ones are Proofpoint, Abnormal Security, and Mimecast.

- **Machine learning algorithms**: Used to analyze email content for suspicious patterns and adapt to emerging threats.

- **Threat intelligence feeds**: Provide real-time updates on known threats to enhance email security measures.

- **Sandboxing techniques**: Open and analyze attachments and links in a secure, isolated environment to detect malicious behavior.

- **Email authentication protocols**: Implement SPF, DKIM, and DMARC to verify the legitimacy of email senders and prevent email spoofing.

- **User awareness and training**: Educate users on recognizing phishing emails, safe email practices, and reporting suspicious emails.

Implementing these email security strategies ensures a robust defense against the sophisticated tactics used in weaponization, safeguarding organizational assets, and maintaining the integrity of email communications.

Threat intelligence and monitoring

Threat intelligence and monitoring are vital components in countering weaponization threats, as they provide real-time insights into potential and active threats, enabling organizations to take proactive measures. Threat intelligence involves gathering, analyzing, and disseminating information about current and emerging threats. This intelligence includes data on attack vectors, malicious IP addresses, malware signatures, and **tactics, techniques, and procedures** (**TTPs**) used by threat actors. By integrating threat intelligence into their security infrastructure, organizations can enhance their ability to detect and respond to threats before they cause significant harm. For example, platforms such as IBM X-Force Exchange and FireEye Helix offer comprehensive threat intelligence feeds that help security teams stay updated on the latest threats and adjust their defenses accordingly.

Continuous monitoring is another critical aspect, involving using tools and techniques to constantly oversee network traffic, endpoint activities, and system logs for signs of suspicious behavior. Solutions such as **security information and event management** (**SIEM**) systems aggregate and analyze data from various sources, providing a centralized view of the security landscape. They generate real-time alerts for anomalies or potential threats, enabling swift investigation and response. By combining SIEM with threat intelligence, organizations can prioritize alerts based on the threat landscape and respond more effectively. Additionally, advanced monitoring tools such as **IDS** and **intrusion prevention systems** (**IPS**) provide an added layer of security by actively scanning for malicious activities and blocking them in real time. This integrated approach ensures a robust defense mechanism, significantly reducing the risk of successful weaponization attacks.

Key techniques include the following:

- **Threat intelligence feeds**: Platforms such as IBM X-Force Exchange and FireEye Helix provide updated data on attack vectors, IP addresses, malware signatures, and TTPs.

- **SIEM**: This aggregates and analyzes data from various sources to provide real-time alerts for anomalies and potential threats.

- **IDSs**: Monitor network traffic for signs of malicious activity and generate alerts when suspicious behavior is detected.

- **IPSs**: Actively scan and block malicious activities in real time, preventing potential breaches.

- **Continuous monitoring**: Constant oversight of network traffic, endpoint activities, and system logs to promptly detect and respond to threats.

- **Integration of threat intelligence and monitoring**: Combining these tools allows for prioritizing alerts based on the current threat landscape, enhancing the effectiveness of response strategies.

Implementing these threat intelligence and monitoring techniques ensures a comprehensive and proactive defense against weaponization threats, helping to safeguard organizational assets and maintain the integrity of the network and systems.

User awareness and training

User awareness and training are critical components in countering weaponization threats, as human error often serves as the weakest link in cybersecurity. Organizations can significantly reduce the risk of successful attacks by educating employees about various cyber threats, such as phishing, social engineering, and malicious attachments. Training programs can help users recognize suspicious emails, avoid clicking on unsafe links, and report potential security incidents promptly. Regular simulations and phishing tests can further reinforce this training, ensuring that employees stay vigilant and can identify real threats. According to a study by the SANS Institute, organizations with robust user awareness programs experience fewer successful phishing attacks and malware infections, demonstrating the tangible benefits of such initiatives.

Moreover, comprehensive user training should include guidelines on best security practices, such as using strong, unique passwords, enabling MFA, and following secure browsing habits. Training should also cover the importance of timely software updates and patch management to prevent the exploitation of known vulnerabilities. By fostering a security-conscious culture, employees become the first line of defense, actively contributing to the organization's overall cybersecurity posture. Resources such as NIST provide frameworks and guidelines for developing effective security training programs, emphasizing the role of continuous education in maintaining robust cybersecurity defenses.

Key techniques include the following:

- **Phishing awareness training**: Educate employees on recognizing and avoiding phishing attempts.
- **Regular simulations and phishing tests**: Conduct exercises to reinforce training and ensure vigilance.
- **Best security practices**: Train employees to use strong, unique passwords, enable MFA, and maintain secure browsing habits.
- **Timely software updates and patch management**: Emphasize the importance of keeping software updated to prevent exploitation of vulnerabilities.
- **Reporting mechanisms**: Establish clear protocols for reporting suspicious emails and potential security incidents.
- **Continuous education**: Implement ongoing training programs to inform employees about the latest threats and security practices.

Implementing these user awareness and training techniques ensures a well-informed and vigilant workforce, significantly enhancing the organization's ability to counter weaponization threats and maintain a strong cybersecurity posture.

Effective strategies to counter weaponization threats are essential for safeguarding organizations against cyberattacks. Implementing rigorous patch management practices that adopt application allowlisting, endpoint and email security, network security measures, user awareness, and training can significantly enhance an organization's cybersecurity posture.

Summary

This chapter delved into the critical phase of the cyber kill chain, where attackers transform gathered intelligence into actionable malicious payloads. The weaponization process involves the meticulous crafting of malware designed to exploit specific vulnerabilities in the target system. This section covered the diverse types of malware and payloads, including viruses, worms, Trojans, ransomware, and spyware, highlighting their unique characteristics and the specific threats they pose.

We also looked at case studies of notable weaponization attacks, such as the infamous NotPetya ransomware attack, demonstrating the real-world impact of these sophisticated payloads. These examples underscore the importance of understanding the vulnerabilities and exploits used in weaponization, which are critical for creating effective defenses. The discussion extends to common vulnerabilities such as unpatched software, misconfigurations, and weak passwords and to common exploits such as buffer overflows, SQL injection, and XSS.

To mitigate these threats, we outlined strategies to counter weaponization threats, emphasizing proactive measures, including patch management to keep software secure; an application allows the listing to control software execution, robust endpoint security, email security, and user awareness training. These strategies form a multi-layered defense against the sophisticated tactics used in weaponization.

With a thorough understanding of how attackers prepare their payloads, we will now transition to the next critical phase in the cyber kill chain: delivery. The next chapter will explore the various methods attackers use to deliver these weaponized payloads to their intended targets, setting the stage for exploitation and further attack stages. Understanding delivery techniques is crucial for implementing effective defenses and stopping attacks before they can execute their malicious actions.

Further reading

Following are some bonus reading materials for you to check out:

- *Analysis of a botnet takeover, providing insights into botnet operations and control mechanisms*: https://research.birmingham.ac.uk/en/publications/your-botnet-is-my-botnet-analysis-of-a-botnet-takeover

- *Overview of the Carbanak APT group's billion-dollar cyberheist targeting banks worldwide*: https://www.infosecinstitute.com/resources/threat-intelligence/carbanak-cybergang-swipes-1-billion-banks/

- *Research on zero-day vulnerabilities, their discovery, and implications for cybersecure*: https://www.zero-day.cz/research/

- *Comprehensive guide on SQL Injection attacks, including prevention techniques and examples*: https://owasp.org/www-community/attacks/SQL_Injection

- *Detailed explanation of Cross-Site Scripting (XSS) attacks, types, and mitigation strategies*: https://owasp.org/www-community/attacks/xss/

- *Exploration of human factors in cybersecurity, focusing on phishing attacks and their impact*: https://www.isaca.org/resources/isaca-journal/issues/2023/volume-4/the-role-of-human-factors-engineering-in-cybersecurity

- *In-depth analysis of the Equifax data breach and the company's response to the incident*: https://krebsonsecurity.com/2017/09/equifax-breach-setting-the-record-straight/

- *Discussion on the challenges of patching systems in the context of the WannaCry ransomware attack*: https://www.eff.org/deeplinks/2017/05/why-patching-problem-makes-us-wannacry

Subscribe to _secpro – the newsletter read by 65,000+ cybersecurity professionals

Want to keep up with the latest cybersecurity threats, defenses, tools, and strategies?

Scan the QR code to subscribe to **_secpro**—the weekly newsletter trusted by 65,000+ cybersecurity professionals who stay informed and ahead of evolving risks.

https://secpro.substack.com

Get This Book's PDF Version and Exclusive Extras

UNLOCK NOW

Scan the QR code (or go to packtpub.com/unlock). Search for this book by name, confirm the edition, and then follow the steps on the page.

Note: Keep your invoice handly. Purchase made directly from Packt don't require one.

4

Delivery

Delivery delves into the intricate process through which cyber adversaries transmit malicious payloads to their intended targets. This phase is pivotal within the cyber kill chain, transitioning from reconnaissance and weaponization to active engagement with the victim. The delivery phase is a critical step in the cyber kill chain, where cybercriminals put their weaponized payloads into action by delivering them to their intended targets. This phase encompasses various **tactics, techniques, and procedures (TTPs)** employed by adversaries to deliver malicious software or exploits to unsuspecting victims. Standard delivery methods include phishing emails, malicious websites, drive-by downloads, and other social engineering strategies.

Real-world incidents, such as the SolarWinds supply chain attack and the Colonial Pipeline ransomware attack, highlight the devastating impact of successful delivery-based cyberattacks. This chapter delves into these case studies, examining the techniques for detecting and blocking delivery attacks. It explores the effectiveness of web security solutions in safeguarding against malicious links and emphasizes the importance of educating users about safe email and web practices. Furthermore, the role of user awareness and training initiatives are discussed as essential components in fortifying defenses against delivery stage attacks. In this chapter, we will cover the following topics:

- Delivery methods
- Recent trends in malware delivery methods
- Evolution of email attachments
- Malware deception methods
- Real-world incidents – delivery-based cyberattacks
- Techniques for detecting and blocking delivery attacks
- Educating users about safe email and web practices
- End-user training initiatives and their importance

Delivery methods

The delivery methods employed by cybercriminals are diverse and sophisticated, each tailored to exploit specific vulnerabilities and deceive targeted victims. Email-based attacks, such as phishing, **business email compromise** (BEC), and spear phishing, are among the most prevalent strategies. **Phishing** involves sending deceptive emails that lure recipients into divulging sensitive information or clicking on malicious links. Conversely, BEC targets businesses by impersonating executives or trusted contacts to authorize fraudulent transactions. Spear phishing takes a more personalized approach, crafting tailored emails that appear credible to specific individuals. Beyond email, drive-by downloads represent another standard method, where unsuspecting users are infected with malware simply by visiting compromised websites. These delivery tactics are designed to bypass traditional security measures, making it essential for organizations and individuals to understand and recognize these threats to defend against them effectively. Following the discussion on delivery methods, we will explore various protection strategies to safeguard against these threats, including email filtering techniques, web security solutions, and user awareness training. Following is the list of various delivery methods:

- **Email-based attacks**: Email-based attacks continue to be a prevalent and effective method for the delivery of weaponized payloads. This concise yet comprehensive discussion explores the key subcategories within this domain, namely phishing, spear phishing, and BEC, providing a deep insight into their mechanisms and implications:

    - **Phishing**: Phishing emails represent a pervasive threat in the cyber landscape. They are carefully designed to impersonate legitimate entities, luring recipients into clicking on malicious links or downloading infected attachments. The success of phishing emails often hinges on exploiting human psychology, leveraging emotions such as fear or curiosity. This manipulation of the human element significantly increases their success rate.

    - **Spear phishing**: Spear phishing takes email-based attacks to a more targeted level. This technique involves tailoring messages to specific individuals or organizations, incorporating highly personalized content that is often difficult to distinguish from legitimate correspondence. Attackers invest time in researching their victims to craft convincing emails, making them appear trustworthy and genuine. The sophistication of spear phishing makes it a potent tool for cybercriminals.

    - **BEC**: BEC attacks involve a more nuanced approach. In these scenarios, attackers impersonate high-level executives or employees within an organization. The goal is to deceive recipients into taking actions that may have far-reaching consequences, such as transferring funds or revealing sensitive information. BEC attacks often exploit the trust and authority associated with these high-ranking individuals, making them particularly effective.

- **Malvertising**: Short for **malicious advertising**, malvertising is a technique where cybercriminals embed malicious code within online advertisements to infect users' devices with malware. These deceptive ads often appear on legitimate websites, making it difficult for users to discern their malicious intent. Cybercriminals exploit vulnerabilities in ad networks to distribute these malicious ads widely. They use fake virus alerts, enticing offers, or mimicking legitimate ads to lure victims into clicking on them. Once clicked, the ads can download and execute malware on the user's device without their knowledge, leading to potential data breaches, system compromise, or financial loss. To mitigate the risks associated with malvertising campaigns, it is crucial to implement ad-blocking software to prevent malicious ads from displaying and protect users from inadvertently clicking on them. Following is an example of malvertising:

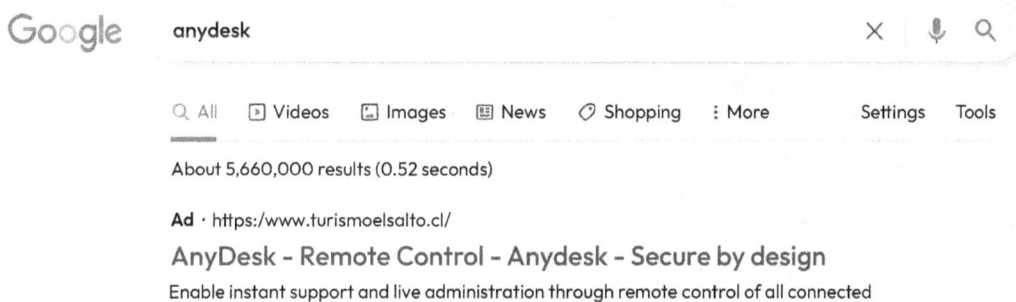

Google   anydesk   × 🎤 🔍

🔍 All   ▶ Videos   🖼 Images   📰 News   ⊘ Shopping   ⋮ More   Settings   Tools

About 5,660,000 results (0.52 seconds)

Ad · https:/www.turismoelsalto.cl/
AnyDesk - Remote Control - Anydesk - Secure by design
Enable instant support and live administration through remote control of all connected

Figure 4.1 – Malvertising on Google Ads

- **SEO poisoning**: SEO poisoning, or **search engine poisoning**, is a tactic where threat actors manipulate search engine rankings to redirect users to malicious websites. Cybercriminals achieve this by injecting malicious code into compromised websites or creating fraudulent sites optimized with trending or popular keywords. These tactics deceive search engines, causing the malicious sites to appear high in search results. Unsuspecting users who click on these results are redirected to pages that can host malware, phishing schemes, or exploit kits to steal sensitive information. The impact of SEO poisoning is significant, as it increases users' exposure to cyber threats and undermines trust in search engine results. To prevent SEO poisoning, it is essential to regularly monitor website integrity, ensure robust cybersecurity measures, and educate users on the risks of clicking on unfamiliar search results. This proactive approach can help safeguard against these malicious activities and protect users from potential harm.

- **Remote monitoring and management (RMM)**: RMM tools have emerged as a significant delivery method for cyber adversaries in 2023. Threat actors have exploited these legitimate tools to establish a stealthy foothold in target systems. They used sophisticated social engineering techniques to trick victims into downloading RMM tools such as ConnectWise,

TeamViewer, AnyDesk, ScreenConnect, and NetSupport Manager installers. By leveraging these legitimate tools, attackers bypass traditional security measures, maintain persistent access, and conduct further malicious activities. This method of using RMM tools as a delivery mechanism underscores the evolving tactics of cybercriminals who exploit trusted software to carry out their operations stealthily and effectively.

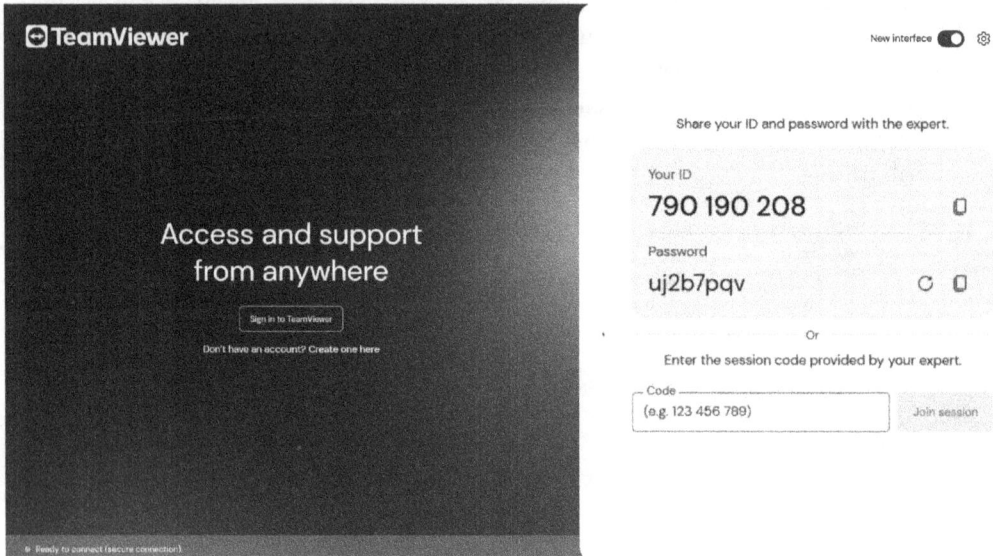

Figure 4.2 – TeamViewer interface

Email-based attacks serve as the initial vector for a wide range of cyberattacks. Phishing emails prey on unsuspecting users, while spear phishing elevates the danger by focusing on specific targets. These attacks are frequently the entry point for the delivery of malware or the compromise of sensitive information, emphasizing the urgency of understanding and mitigating these threats.

Invoice_no-WDE346TS

Dear User,

We notice unauthorized transactions from your PayPal account If You do not make this transaction then kindly call us for a query of this order otherwise, your USD 589.23 has been charged today. If you want to cancel this order, give us a call on Helpline Number: +1 (832) 539 4235

Transaction date- AUG-06-2024

Your

BITCOIN EXCHANGE

Amount USD 589.32

Thank you for choosing our service.
If you do not authorize this transaction or if you didn't make this purchase don't hesitate to get in touch with our support team at the earliest to cancel it or issue a refund
Thank You
Team PayPal

Terms & Conditions

You got a money request for an unpaid service. This request remains unpaid until you pay this Due. Pay via your Card/PayPal and clear this amount immediately.

The amount will be auto-debited by your given credit/debit or bank account within 24 Hours.

Helpline Number +1 (832) 539 4235

Copyright @ 2024 PayPal. All rights reserved

Figure 4.3 – Phishing email

Email-based attacks, including phishing, spear phishing, and BEC, represent a persistent and evolving cybersecurity challenge. Their success in exploiting human vulnerabilities and organizational trust underscores the need for robust defenses and awareness. As cyber threats continue to evolve, researchers and practitioners must remain vigilant in their efforts to combat these insidious forms of cybercrime.

- **Drive-by Downloads**: Drive-by downloads are a prevalent form of cyber-attack where cybercriminals exploit vulnerabilities within web browsers or associated plugins to discreetly deliver malware to unsuspecting users. In these attacks, victims can become infected without any active engagement, merely by visiting a compromised website.

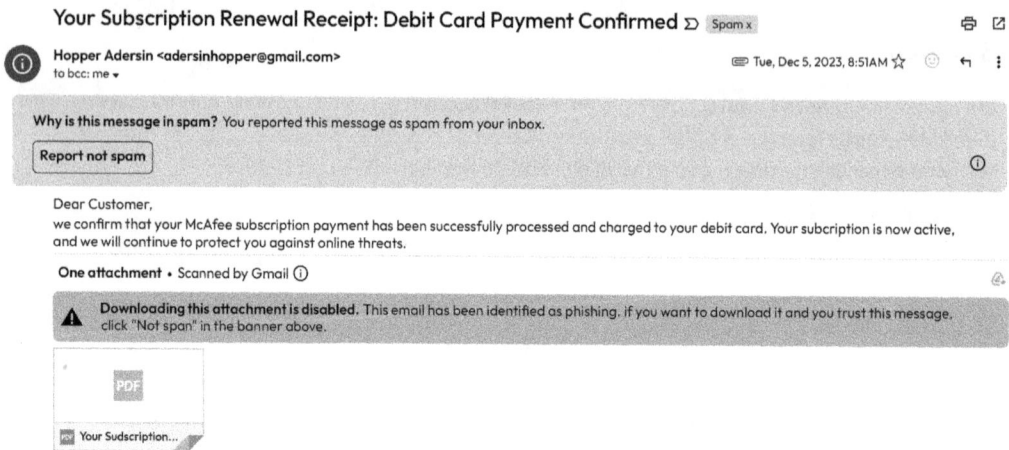

Figure 4.4 – Phishing email with an attachment

These attacks hinge on the exploitation of security weaknesses in web browsers, plugins, or web applications. Cybercriminals adeptly leverage these vulnerabilities to introduce malicious software onto the user's device. The most alarming aspect is that users may remain entirely unaware of this intrusion, as it occurs silently in the background.

The success of drive-by download attacks is often contingent on the exploitation of unpatched software or the utilization of zero-day vulnerabilities. Unpatched software refers to applications or systems that have not been updated to address known security flaws. In such cases, cybercriminals take advantage of these known weaknesses. On the other hand, zero-day vulnerabilities are those that were previously undisclosed and unaddressed by the software developer. These vulnerabilities hold immense allure for cybercriminals, as they provide a unique advantage by exploiting an unknown flaw.

Email-based attacks, including phishing and spear phishing, prey on human vulnerabilities, while drive-by downloads take advantage of software weaknesses. Understanding these delivery methods is crucial for organizations and individuals to bolster their defenses against cyber threats.

Recent trends in malware delivery methods

In the ever-evolving landscape of cybersecurity, understanding the latest trends in malware delivery is crucial for developing robust defense strategies. As cybercriminals refine their tactics, their methods to infiltrate systems become more sophisticated. This section delves into the nuanced techniques attackers use, providing practical insights to help protect against these evolving threats [see 1 in *Further Readings*]:

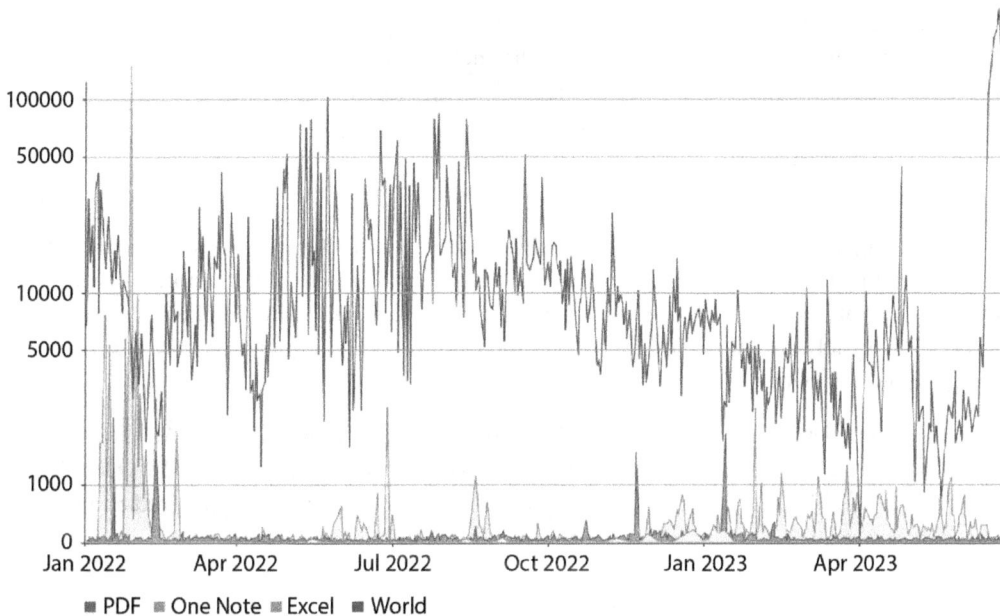

Figure 4.5 – File types used as malicious attachments since 2022

Evolution of email attachments

Email remains a primary vector for malware delivery. Despite being a decades-old technique, email attachments continue to thrive due to their effectiveness in bypassing security measures and exploiting user trust. Traditionally, file types such as Excel, RTF, CAB, and various compressed formats dominated this space. However, recent data from VirusTotal indicates a shift toward new file formats that offer better evasion capabilities.

The rise of OneNote and ISO files

One of the key trends we have observed when it comes to malware delivery through emails is the rise of OneNote and ISO files. Let's have a look.

OneNote

In 2023, attackers increasingly adopted Microsoft OneNote as a reliable alternative to macros in other Office products. OneNote's flexibility allows embedding malicious URLs and various scripting formats, including JavaScript, PowerShell, Visual Basic Script, and Windows Script. Initially, **antivirus (AV)** products struggled to detect OneNote-based malware, giving attackers a window of opportunity. This format has distributed multiple malware families, including Qakbot, IcedID, Emotet, AsyncRAT, and Redline Stealer.

One notable trend is the use of OneNote files as password-encrypted email attachments. These files often execute embedded scripts that download additional malware. The success of these methods is evident in the significant increase in OneNote-related submissions to VirusTotal in 2023, indicating its growing popularity among cybercriminals.

ISO files

Another emerging trend is using ISO files for malware distribution. Often heavily compressed, ISO files are challenging for security solutions to scan effectively. In 2023, there was a noticeable uptick in suspicious ISO files, which were used to distribute various malware, including LockBit, DarkBit, Quakbot, and AsyncRAT. These files are frequently disguised as legitimate software installation packages, further complicating detection efforts.

The data shows smaller ISO files (less than 15 MB) are more likely to be malicious. Attackers often use intermediate archives, such as compressed files, to distribute ISO files via email. This technique leverages ISO files' ability to evade detection by AV engines, making them a preferred choice for malware delivery in 2023.

JavaScript and HTML in phishing campaigns

Alongside OneNote and ISO files, there has been a significant increase in the use of JavaScript distributed along with HTML for phishing attacks. These elaborate phishing campaigns aim to steal victims' credentials and other sensitive information. This trend highlights the attackers' shift toward leveraging web technologies to craft more convincing and compelling social engineering attacks.

Decline of traditional file types

While new formats gain popularity, traditional file types such as Excel, RTF, and Word are seeing a relative decline in usage as malicious attachments. However, these formats are only partially obsolete. For instance, there were notable peaks in suspicious Excel and Word files in mid-2023, suggesting opportunistic usage based on specific campaigns.

The rotation of formats from traditional to emerging ones underscores the attackers' continuous efforts to bypass security measures and exploit new vulnerabilities. By understanding these emerging trends in malware delivery, cybersecurity professionals can better anticipate and mitigate the risks posed by increasingly sophisticated cyber threats. This proactive approach is essential for maintaining robust defenses despite evolving attack methodologies.

Malware deception methods

Understanding the deceptive methods cyber adversaries use is crucial for enhancing defense mechanisms. **Malware deception** is a strategic approach attackers use to infiltrate systems and evade detection, often exploiting the implicit trust in legitimate entities. This section delves into cybercriminals' various techniques and strategies to deceive their targets and achieve their malicious objectives.

Abuse of trust

One of the fundamental strategies in malware deception is the abuse of trust. Attackers leverage users' inherent trust in legitimate domains, software, and certificates to distribute malware. This abuse can manifest in several ways, such as the following.

Distribution through legitimate domains

In the digital age, robust web security solutions are essential to protect users from malicious links and other online threats. In the absence of robust security, attackers can use domains as gateways to carry out the attack:

- **Technique**: Attackers often use well-known and reputable domains to distribute their malware, which helps them bypass traditional security defenses, such as domain/IP-based firewalls.

- **Advantage**: By using legitimate domains, attackers can avoid needing dedicated infrastructure that can be easily traced and taken down.

- **Example**: Cyber adversaries have been known to distribute malware through top domains such as `amazonaws.com`, Azure, and `mediafire.com`, exploiting users' trust in these platforms.

Malware signed with legitimate certificates

Malware signed with legitimate certificates uses authentic digital signatures to appear trustworthy, allowing it to bypass traditional security measures. This tactic exploits the inherent trust in legitimate certificates, making detection and prevention challenging:

- **Technique**: Malware is signed with stolen or misused legitimate signing certificates, making it appear from a trusted source.

- **Advantage**: Signed malware is often considered safe by operating systems and security solutions, thereby increasing the likelihood of successful infiltration.

- **Example**: The infamous Lapsus$ group stole Nvidia's signing certificates and used them to sign malicious samples, effectively masquerading as legitimate Nvidia software.

Mimicking legitimate software

Deceptive malware frequently mimics the appearance and behavior of legitimate software to trick users into executing it. This method is particularly effective in social engineering attacks.

Visually similar icons

To manipulate user behavior, attackers often disguise malware using icons that closely resemble well-known applications:

- **Technique**: Malware is disguised using icons that are visually similar to those of popular legitimate applications.

- **Advantage**: Users are more likely to trust and execute files that appear familiar and legitimate.

- **Example**: Malware mimicking the icons of applications such as Skype, Adobe Acrobat, and VLC has been widely observed, exploiting the high trust and frequent usage of these applications.

Packaging with legitimate software

In another common technique, malware is bundled with genuine software, making it harder for users to distinguish between safe and malicious files:

- **Technique**: Malware is bundled with legitimate software installers within the same compressed file or embedded as a resource.

- **Advantage**: This approach can deceive users into believing they are installing a legitimate application while the malware is executed in the background.

- **Example**: Attackers have distributed malware alongside legitimate installers for software such as Google Chrome, Zoom, and Proton VPN, ensuring the malicious code runs unnoticed.

Real-world examples

Understanding these deceptive methods through real-world examples provides practical insights for cybersecurity professionals.

Mimicking websites with favicons

Attackers frequently replicate favicons from trusted websites to lure users into visiting malicious domains. This technique leverages the trust users place in popular platforms:

- **Technique**: Attackers use favicons (small website icons) similar to legitimate websites to deceive users into visiting malicious sites.

- **Example**: Popular platforms such as WhatsApp, Instagram, and Amazon have been targets for favicon mimicry, with attackers creating phishing sites that appear visually identical to these trusted brands.

Execution parents

By embedding malware within legitimate software's execution flow, attackers can hide malicious actions, making it significantly more challenging to detect and analyze:

- **Technique**: Malware leverages legitimate software execution flows to mask its activities. By executing alongside or within legitimate software, malware can operate covertly.

- **Example**: Instances where malware is executed alongside legitimate installers, such as those for Malwarebytes or Firefox, make detection and attribution more challenging.

The effectiveness of malware deception methods underscores the importance of continuous vigilance and adaptation in cybersecurity practices. Understanding these tactics is vital for professionals and organizations to enhance threat intelligence and implement robust defense mechanisms. Regularly updating and monitoring malware trends can help anticipate new deceptive strategies and adapt defenses accordingly. Implementing automated systems to detect anomalies in software certificates, domain usage, and software behavior can preemptively identify and mitigate threats. Educating users about the risks associated with downloading and executing software from unverified sources can reduce the success rate of social engineering attacks while encouraging the verification of digital signatures and the scrutiny of software origins before installation further strengthens security postures. Adopting a zero-trust model, where no entity is inherently trusted, can significantly mitigate the risks posed by deceptive malware, ensuring that all internal or external interactions are continuously verified, thereby reducing the likelihood of successful infiltrations.

The evolving tactics of cyber adversaries in malware deception highlight the critical need for advanced threat intelligence and proactive defense strategies. By understanding and anticipating these deceptive methods, cybersecurity professionals and organizations can enhance their defenses and mitigate the risks of sophisticated cyber threats. The insights and examples in this section aim to empower readers with actionable knowledge to effectively counteract these evolving threats.

Real-world incidents – delivery-based cyberattacks

Cyberattacks have become increasingly sophisticated and prevalent in recent years, with delivery-based cyberattacks being a prominent threat vector. In this section, we will examine real-world incidents of delivery-based cyberattacks, highlighting their impact on organizations and the valuable lessons that can be drawn from these cases.

Case Study 1 – SolarWinds supply chain attack

The **SolarWinds supply chain attack,** discovered in late 2020, was a highly sophisticated cyber espionage operation. Attackers compromised SolarWinds' software updates to deliver malicious payloads to thousands of organizations.

This supply chain attack compromised numerous government agencies and private sector organizations. It granted attackers access to sensitive data and intelligence, raising concerns about national security.

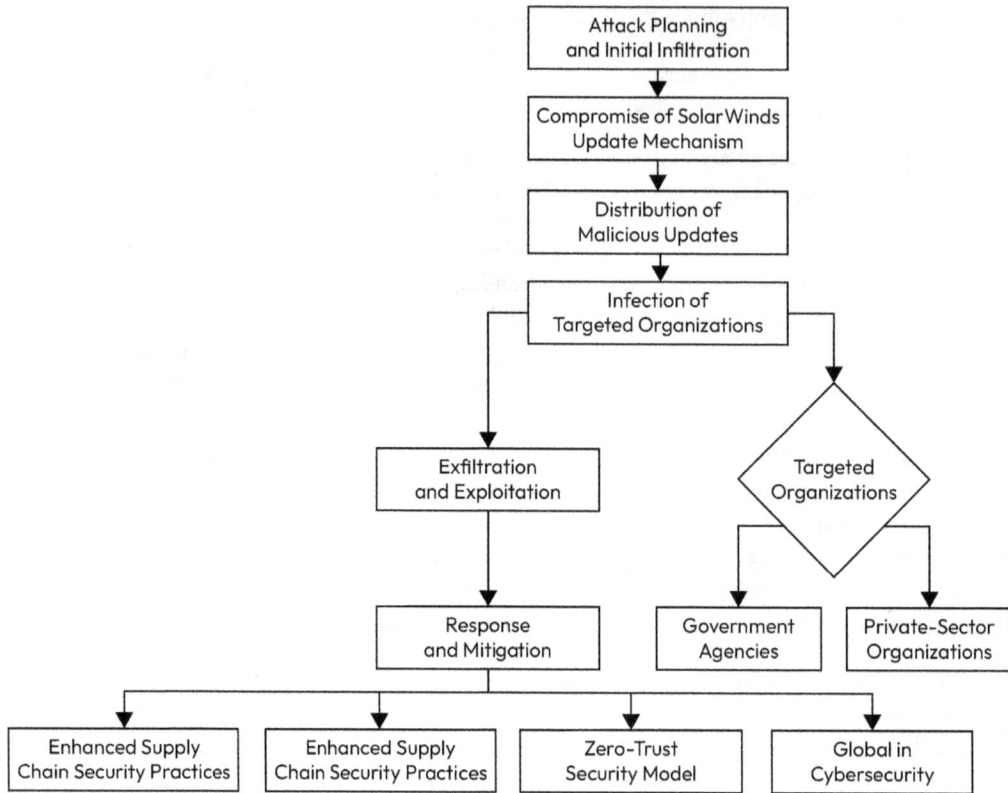

Figure 4.6 – Solarwinds overview

The SolarWinds incident underscored the need for enhanced supply chain security practices, including rigorous vetting of software vendors and their security practices. It also highlighted the importance of continuous monitoring and threat detection to identify unusual behavior.

The SolarWinds supply chain attack, discovered in late 2020, stands as a landmark case of a highly sophisticated and audacious cyber espionage operation. In this attack, malicious actors compromised SolarWinds, a trusted software provider, to deliver malicious payloads via the company's legitimate software updates to thousands of organizations, including some of the most sensitive government agencies.

The SolarWinds supply chain attack had far-reaching consequences, compromising numerous government agencies and private sector organizations. SolarWinds software was a staple in the IT infrastructure of numerous entities, making it an ideal vector for infiltration. The attack granted malicious actors access to sensitive data and intelligence, raising concerns about national security.

SolarWinds' Orion platform, a widely used network management tool, was the primary target of the breach. By injecting a malicious code update into the software, attackers succeeded in infiltrating the systems of SolarWinds' clients. This compromise enabled the unauthorized access of networks, potentially granting the perpetrators unrestricted access to an extensive range of classified information.

The victims of this attack included agencies and organizations with significant responsibilities in the **United States** (**US**), such as the Departments of Defense, State, and Homeland Security, as well as the Treasury, Commerce, and Energy Departments. Internationally, it affected other government entities and private sector organizations, making the scale of the attack unprecedented. The breach was a glaring indication of the potential vulnerability of the supply chain, where compromising a single trusted vendor can impact thousands of interconnected organizations.

The SolarWinds incident has cast a spotlight on the vital importance of enhanced supply chain security practices. A supply chain attack takes advantage of trust – organizations trust their suppliers and the integrity of the software they provide. The attacker leveraged this trust to infiltrate the client's networks and systems.

Organizations should conduct thorough vetting and due diligence when selecting software vendors. This includes assessing their cybersecurity practices, vulnerability management, and overall security posture. Evaluating the vendor's security practices can prevent compromised software from entering the organization's ecosystem.

Continuous monitoring of network traffic and system behavior is crucial. By analyzing network data for anomalies and unusual behavior, organizations can quickly detect and respond to any potential breaches. Suspicious activities should trigger immediate investigations.

The attack reinforced the need for a zero-trust security model, where trust is never assumed, even within the organization's network. This model includes strict access controls, the principle of least privilege, and micro-segmentation to limit lateral movement.

A robust incident response plan is essential. Organizations should have a well-defined and regularly updated incident response plan outlining the actions to be taken in case of a security breach. This includes coordination with law enforcement and information sharing within the cybersecurity community.

The attack highlighted the need for international cooperation in addressing cyber threats. Cyberattacks often transcend national boundaries, necessitating collaborative efforts among nations to identify and apprehend threat actors.

Organizations responsible for sensitive data or critical infrastructure need to adopt enhanced security measures, including advanced threat detection, encryption, and enhanced authentication methods.

The SolarWinds supply chain attack serves as a critical case study that highlights the vulnerabilities in supply chain security. Organizations must be proactive in enhancing their security practices, including scrutinizing their software vendors, implementing robust continuous monitoring, and adopting a zero-trust security model to protect against potential supply chain attacks.

This case underscores that in the ever-evolving landscape of cyber threats, the ability to adapt and improve security practices is paramount to safeguarding national security and the integrity of sensitive data and information. Cybersecurity practitioners and policymakers must draw lessons from this attack to better protect organizations and critical infrastructure in the future.

Case study 2 – Colonial Pipeline ransomware attack

The **Colonial Pipeline ransomware attack**, which unfolded in May 2021, stands as a stark reminder of the vulnerabilities that underlie critical infrastructure in the US. In this case study, we delve into the specifics of the attack, the profound impact it had on the south-eastern US, and the lessons that can be drawn from this significant incident.

The Colonial Pipeline company operates one of the largest fuel pipelines in the US, spanning from the Gulf Coast to the East Coast. On a fateful day in May 2021, a criminal hacking group named Dark Side targeted Colonial Pipeline with a ransomware attack. Dark Side employed a relatively new but highly effective strain of ransomware to infiltrate the company's systems. This ransomware was engineered for encryption and data exfiltration, making it a potent threat.

The impact of the Colonial Pipeline ransomware attack was immediate and severe. The attackers encrypted critical systems within the company, rendering them inaccessible and causing significant disruption to its operations. However, what made this attack particularly alarming was its cascading effect on the fuel supply chain in the south-eastern US.

As Colonial Pipeline serves a significant portion of the south-eastern US, its disruption sent shockwaves through the region. Gasoline, diesel, and jet fuel supplies were interrupted, leading to panic buying by consumers. Long lines formed at gas stations and many were left without access to fuel for their daily needs.

The economic consequences were substantial. The disruption of fuel supplies had a ripple effect on various industries, from transportation and logistics to manufacturing and agriculture. Businesses reliant on timely fuel deliveries suffered significant losses.

The attack raised concerns about national security, highlighting the vulnerability of critical infrastructure to cyber threats. The pipeline's importance to the nation's energy supply made it a tempting target, and the incident underscored the potential impact of cyberattacks on daily life.

The Colonial Pipeline incident provides several crucial lessons for organizations and governments in the context of cybersecurity and critical infrastructure protection.

The attack highlighted the urgent need for securing critical infrastructure, including energy, water, and transportation systems. Organizations operating these assets must invest in robust cybersecurity measures to prevent and mitigate cyber threats effectively.

The incident underscored the significance of having well-defined incident response plans in place. A rapid and coordinated response is crucial in minimizing the impact of cyberattacks on critical infrastructure. Organizations should regularly test their response plans to ensure their effectiveness.

The Colonial Pipeline attack shed light on the necessity for closer collaboration between the public and private sectors in addressing cyber threats. Critical infrastructure companies should work hand in hand with government agencies and law enforcement to enhance cybersecurity and incident response capabilities.

Ransomware attacks, such as the one targeting Colonial Pipeline, continue to be a pervasive threat. Organizations should invest in robust backup and recovery solutions, train employees to recognize phishing attempts, and ensure that software and systems are up to date with security patches.

This attack highlighted the importance of sharing threat intelligence among organizations to identify and mitigate cyber threats promptly. A proactive approach to threat intelligence can help organizations stay ahead of potential attacks.

The impact on fuel supplies and daily life in the south-eastern US underscores the critical need for enhanced cybersecurity measures, incident response planning, public-private cooperation, and the ongoing battle against ransomware threats.

Defense in depth

The defense-in-depth approach in cybersecurity is of paramount importance, as it involves multiple layers of protection to safeguard an organization's assets from various threats. These layers, each serving a unique purpose, include the following:

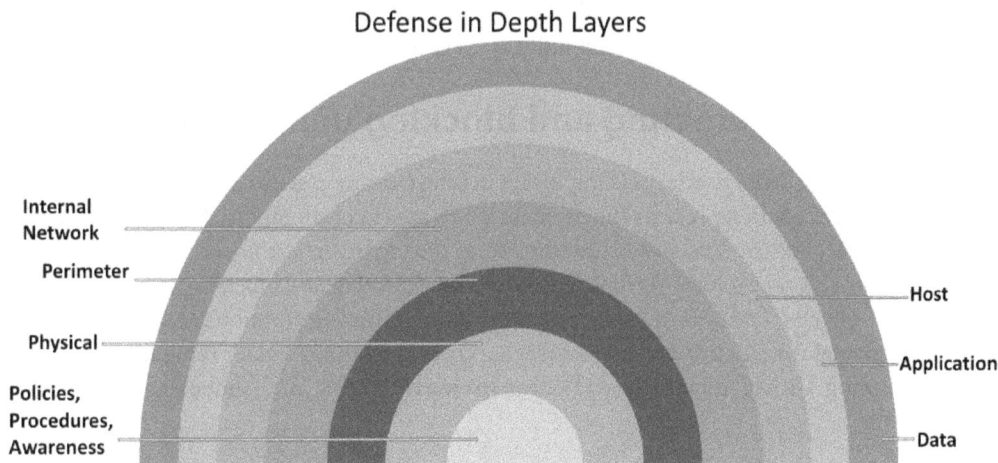

Figure 4.7 – Defense in depth layers

- **Physical layer:** This layer plays a crucial role in cybersecurity by focusing on securing the physical environment through locks, fences, and security guards. These measures are instrumental in preventing unauthorized physical access to facilities and enhancing overall security posture.

- **Policies, procedures, and awareness**: This layer emphasizes the importance of human factors in security, incorporating strong passwords, well-defined security policies, data classification, and training to raise employees' awareness.

- **Perimeter layer**: At this level, security mechanisms like firewalls, VPNs, and packet filters are deployed to control traffic entering and leaving the network, forming the first defense against external threats.

- **Internal network layer**: This, within the internal network, is instrumental in cybersecurity. It uses firewalls, intrusion detection systems, and encryption to monitor and protect internal traffic, providing a robust barrier to prevent unauthorized access or movement within the network.

- **Host layer**: This layer focuses on securing individual devices, including platform operating systems, applying patches, and utilizing malware protection to defend against attacks targeting specific hosts.

- **Application layer**: Security at the application level includes **single sign-on** (**SSO**), authentication, and authorization mechanisms to control access to application resources and ensure that only authorized users can interact with sensitive applications.

- **Data layer**: The innermost layer protects the core data through database, content filtering, and message security, ensuring that sensitive information remains confidential, integral, and available only to authorized parties.

These layers combine to create a robust, multi-faceted security strategy that defends against a wide range of cyber threats by addressing vulnerabilities at every level of an organization's infrastructure.

Techniques for detecting and blocking delivery attacks

Detecting and blocking delivery attacks is important in safeguarding systems and data. The delivery phase of the cyber kill chain represents the moment when attackers attempt to deliver weaponized malware to the target system. Effective countermeasures during this phase are crucial to thwarting cyber threats. This section explores various techniques for detecting and blocking delivery attacks, including email filtering and web security solutions. Email filtering solutions and web security measures play pivotal roles in this defense, leveraging content analysis, sender reputation analysis, attachment scanning, URL filtering, **web application firewalls** (**WAFs**), and sandboxing to safeguard systems and data.

Email filtering – a frontline defense

Email filtering is a technique used to manage and organize incoming email messages by automatically processing them according to predefined criteria. This process involves scanning and categorizing emails to identify and block spam, phishing attempts, and malicious content before they reach the recipient's inbox. Email filtering systems use a combination of methods, including keyword analysis, blacklists, whitelists, and advanced algorithms that detect patterns indicative of unwanted or harmful emails. By employing these techniques, email filtering helps reduce the volume of junk mail, enhance security, and ensure that legitimate messages are delivered promptly. It is an essential tool for both individuals and organizations to maintain efficient communication and protect against various email-based threats. Let's dive into the different types of filtering:

- **Content-based filtering**: Content-based email filtering is a foundational approach in the cybersecurity arsenal. It involves a meticulous examination of the content within emails, encompassing not only the text but also attachments and embedded links. The primary objective here is to identify and neutralize potential threats by leveraging predefined patterns and heuristics. This method is highly effective in recognizing known malware signatures and detecting suspicious email behavior. Content-based filtering is a critical initial layer of protection that helps in identifying and blocking malicious content before it even reaches the recipient. This technique is grounded in the analysis of email content to spot telltale signs of malware. It relies on established databases of known malware signatures and suspicious patterns. For instance, it can detect patterns in email attachments that match known viruses or malware. In the realm of content-based filtering, it's essential to emphasize that accuracy is paramount. It's critical to strike a balance between identifying threats and minimizing false positives, ensuring that legitimate emails are not inadvertently blocked.

- **Sender reputation analysis**: In today's digital landscape, assessing the trustworthiness of email senders is an imperative aspect of email filtering. Sender reputation analysis scrutinizes the sender's historical data and their adherence to sender authentication mechanisms. This scrutiny is crucial for identifying and blocking emails sent by suspicious or unauthenticated senders. By maintaining a watchful eye on sender reputations, organizations can significantly reduce the risk of delivery attacks. Sender reputation is often determined based on factors such as the sender's previous email history and compliance with authentication protocols such as **Sender Policy Framework** (**SPF**) and **DomainKeys Identified Mail** (**DKIM**). Suspicious or unauthenticated senders are flagged or outright blocked to safeguard the organization's email ecosystem. This approach mitigates phishing attacks and prevents the distribution of malicious content from unreliable sources.

- **Attachment scanning**: The inclusion of attachment scanning is a non-negotiable component of email filtering solutions. This method involves a thorough examination of email attachments to uncover any hidden, malicious payloads. Modern solutions go beyond conventional scanning and employ sandboxing, an advanced technique that executes email attachments in isolated environments. This controlled environment allows for in-depth analysis of the attachment's behavior without jeopardizing the security of the recipient's system. Sandboxing is an effective strategy because it enables the assessment of an attachment's actions when opened. It provides a real-world testing ground to gauge the behavior of potentially dangerous files. If the attachment exhibits any malicious behavior, it can be quarantined or blocked, thus preventing potential breaches.

- **Behavioral analysis**: As cyber threats continue to evolve, advanced email security solutions have incorporated behavioral analysis as a proactive defense mechanism. This technique focuses on identifying anomalies in email behavior, including sender actions and patterns. For instance, it can detect unusual sender behaviors, such as the rapid sending of multiple emails or sudden changes in sending patterns, which may indicate a compromised email account. Behavioral analysis is crucial because it extends beyond static, pattern-based detection. It considers the dynamics of email communication, making it highly adaptive to emerging threats. By identifying these anomalies, organizations can respond swiftly to mitigate the risks associated with compromised accounts and insider threats. This approach significantly enhances the overall security posture of email systems.

These advanced email filtering methods provide a multi-layered approach to protect against a wide range of email-borne threats, from malware and phishing attempts to suspicious sender behavior. Implementing these techniques as part of an organization's cybersecurity strategy is instrumental in safeguarding sensitive information and ensuring the integrity of email communications.

Web security solutions – safeguarding against malicious links

In the digital age, where online threats are increasingly sophisticated, robust web security solutions are essential to protect users from malicious links. These solutions encompass a range of technologies and practices designed to detect, block, and mitigate the risks posed by harmful web content. URL filtering, for instance, prevents access to known dangerous websites by analyzing and categorizing URLs based on their threat levels. WAFs provide an additional layer of defense by actively monitoring and filtering HTTP traffic between a web application and the internet, blocking malicious traffic in real time before it can cause harm. Sandbox environments and file analysis tools enhance security by isolating and examining suspicious files and links in a controlled setting, ensuring they are safe before allowing access. These web security measures play a crucial role in safeguarding against cyber threats and protecting individuals and organizations from the ever-evolving landscape of online attacks:

- **URL filtering**: URL filtering, often considered the first line of defense in web security, is a complex system that operates on several layers, which are described as follows:

  - **Blacklisting and whitelisting**: Beyond maintaining databases of known malicious domains and URLs, URL filtering also employs blacklists and whitelists. Blacklists consist of websites that are known to be harmful, while whitelists are made up of trusted websites. These lists are continuously updated based on emerging threats and user feedback. Researchers at Stanford University highlighted the dynamic nature of URL filtering with constantly evolving blacklists and whitelists.

  - **Machine learning integration**: URL filtering has embraced machine learning techniques to identify new threats. Machine learning models analyze not only the URL itself but also factors such as the website's structure and content, enabling the identification of previously unknown malicious links.

- **WAFs**: WAFs are at the forefront of protecting web applications from a wide range of attacks. To provide a deeper insight into their operations, let's look at the following:

  - **Behavioral analysis**: Modern WAFs scrutinize web traffic in real time. This technique observes user interactions, patterns of access, and request frequencies to detect anomalies and potential threats. By continuously adapting to changing attack strategies, WAFs remain effective against evolving threats. Research by experts at institutions such as Carnegie Mellon University has demonstrated the effectiveness of behavioral analysis in identifying web application attacks in real time, highlighting its role in enhancing the security of web applications against sophisticated cyber threats.

  - **Machine learning and anomaly detection**: WAFs are now incorporating machine learning algorithms for anomaly detection. These algorithms can identify deviations from expected behavior, which is particularly important in recognizing zero-day attacks where attack patterns are not yet known. A comprehensive report published by the **Open Web Application Security Project** (**OWASP**) explores the application of machine learning and anomaly detection techniques in modern WAFs and their contribution to web application security.

- **Sandboxing and file analysis**: Sandboxing and file analysis have evolved significantly, employing advanced methods for an in-depth examination of potential threats:

  - **Micro-virtualization**: Sandboxing now often employs micro-virtualization, where each process is executed in its own micro **virtual machine** (**VM**). This level of isolation ensures that even if a malicious file manages to breach the sandbox, it cannot harm the host system.

- **Content Disarm and Reconstruction (CDR)**: File analysis techniques have expanded to include CDR, a method that involves disassembling and reconstructing files to remove potential threats. This approach is particularly useful in dealing with complex file-based attacks, as it ensures that only sanitized content is allowed through. A report by Trend Micro, a prominent cybersecurity company, outlines the significance of CDR in file analysis and how it effectively neutralizes file-based threats.

The world of advanced web security solutions is multifaceted and continuously evolving. URL filtering, WAFs, and sandboxing / file analysis have become highly sophisticated, integrating machine learning, behavioral analysis, and advanced isolation techniques.

Effective detection and blocking of delivery attacks are essential components of a robust cybersecurity strategy. Email filtering and web security solutions are fundamental tools in this regard, helping organizations proactively defend against malicious attempts to deliver weaponized malware.

Educating users about safe email and web practices

Educating users about safe email and web practices is crucial in mitigating the risk of cyberattacks. User awareness plays a pivotal role in recognizing and responding to these threats. This section explores various aspects of user education, including understanding delivery-based attacks, recognizing phishing attempts, and the importance of regular software updates.

User awareness plays a pivotal role in defending against the delivery stage of the cyber kill chain, where cyber adversaries actively attempt to transmit malicious payloads to their targets. This phase often exploits human vulnerabilities through tactics such as phishing emails, malicious websites, and social engineering schemes. By fostering a culture of cyber vigilance, organizations can empower their employees to recognize and respond to these threats effectively. Training programs that educate users about the latest attack vectors, safe email, web practices, and verifying suspicious communications are essential. Awareness initiatives reduce the likelihood of successful attacks and enhance an organization's overall security posture. Ultimately, our informed and vigilant user base is not just a defense; they're the first line of defense in mitigating the risks associated with the delivery stage of cyberattacks, and they're committed to it. The following points highlight the critical aspects of user awareness in the context of delivery-based cyberattacks, emphasizing the significance of understanding these threats, recognizing phishing attempts, and maintaining regular software updates:

- **Understanding delivery-based attacks**: Delivery-based attacks, including phishing, are on the rise and pose a significant threat to individuals and organizations. These attacks often leverage psychological manipulation, social engineering, and impersonation to deceive users. By raising awareness about the existence and methods of these attacks, individuals can become more vigilant and less susceptible to falling victim to such schemes. According to the Verizon **Data Breach Investigations Report (DBIR)** for 2021, phishing was implicated in a staggering 36% of all data breaches. This statistic highlights the prevalence of phishing attacks and their devastating impact on data security.

- **Recognizing phishing attempts**: Phishing emails are a common vehicle for delivery-based cyberattacks. Users need to acquire the knowledge and skills necessary to identify these deceptive emails effectively. Phishing attempts often exhibit several red flags, including suspicious sender addresses, misspelled URLs, and requests for sensitive information, such as login credentials or financial data. By educating users on these indicators, they can become more adept at recognizing phishing attempts. A study conducted by the **Anti-Phishing Working Group (APWG)** further emphasizes the need for improved user recognition of phishing attempts. The study revealed that the average phishing site's uptime was only 2.5 days during Q3 2021, indicating that these sites are frequently taken down. This underscores the critical role of users in spotting and reporting phishing attacks promptly.

- **Importance of regular updates**: Keeping software, including email clients and web browsers, up to date is a fundamental aspect of cybersecurity. Users should be well versed in the significance of timely software updates, as attackers often target unpatched vulnerabilities. Failure to update software can leave systems vulnerable to exploitation by cybercriminals. A notable real-world example of the consequences of failing to update software is the Hafnium attacks in early 2021. Attackers exploited unpatched vulnerabilities in the Microsoft Exchange Server, leading to widespread breaches and data compromises. This incident underscores the tangible risks associated with neglecting software updates.

End-user training initiatives and their importance

In the ongoing battle against cyber threats, employee and user training initiatives are crucial components of a comprehensive cybersecurity strategy. These initiatives are designed to educate and equip users with the knowledge and skills necessary to recognize and respond to potential cyber threats effectively. Given that cybercriminals often exploit human error and lack of awareness through tactics such as phishing, social engineering, and malicious downloads, training programs play a vital role in reducing these vulnerabilities. By fostering a culture of cyber awareness, organizations can significantly enhance their defense mechanisms. Regular training sessions, simulated attack scenarios, and updated cybersecurity guidelines ensure that employees remain vigilant and informed about the latest threat vectors and best practices. Ultimately, investing in user training initiatives mitigates the risk of successful cyberattacks and strengthens the organization's overall security posture, creating a more resilient digital environment. Here are some of the methods to consider:

- **Simulated phishing exercises**: Simulated phishing exercises are a vital component of user education in cybersecurity. These exercises involve creating scenarios that mimic real-world phishing attempts. Employees receive mock phishing emails, and their responses are monitored. Here's a more in-depth look at this training method:

  - **Organizations design diverse phishing scenarios**: Simulated exercises should encompass a wide range of phishing tactics, from deceptive financial requests to emails purportedly from trusted sources. By exposing employees to a variety of threats, they become better equipped to recognize and respond to different types of phishing attacks.

  - **Feedback and learning**: After employees interact with these simulated phishing emails, immediate feedback is provided. They learn about the indicators of a phishing attempt they missed and understand the consequences of their actions. This immediate feedback mechanism reinforces awareness and reduces the likelihood of falling for real phishing attacks.

  - **Real-world training**: Simulated exercises mirror real-world scenarios, enabling users to develop practical skills. These training sessions familiarize users with the latest phishing techniques and tactics, making them more resilient against evolving threats.

  - **Long-term impact**: The long-term impact of simulated phishing exercises is evident in reduced click rates on phishing emails. Over time, organizations observe a significant decline in employee susceptibility to phishing, contributing to a more secure environment.

- **Security awareness programs**: Security awareness programs play a central role in educating users about safe email and web practices. These programs go beyond phishing simulations and cover various aspects of cybersecurity. Here's an in-depth exploration of this training initiative:

  - **Customized programs**: Security awareness programs should be tailored to the specific needs of the organization. They consider the industry, organizational culture, and the unique threats the organization faces. For instance, a financial institution's security awareness program may emphasize financial fraud prevention, while a healthcare organization's program might focus on safeguarding patient data.

  - **Comprehensive curriculum**: These programs encompass a wide range of topics, including data protection, password management, secure browsing, and mobile device security. The curriculum is regularly updated to address emerging threats.

  - **Employee engagement**: Engaging employees is key to the success of security awareness programs. Gamification, quizzes, and interactive training modules make learning about cybersecurity enjoyable and memorable. Moreover, organizations often reward employees for actively participating in these programs, fostering a culture of security.

  - **Continuous learning**: Security awareness is an ongoing process. These programs promote continuous learning through periodic training sessions and updates on the latest cybersecurity trends and threats.

  - **Evaluation and metrics**: Organizations measure the success of security awareness programs through metrics such as the reduction in security incidents, improved employee awareness, and reduced response times to security incidents.

- **Multi-factor authentication (MFA) education**: Educating users on the benefits and proper usage of MFA is essential in enhancing account security. Here's a comprehensive overview of MFA education:

  - **Benefits of MFA**: Users need to understand that MFA adds an extra layer of security by requiring them to provide multiple forms of verification before accessing their accounts. This significantly reduces the risk of unauthorized access, even if passwords are compromised.

  - **Authentication methods**: MFA involves various authentication methods, including something the user knows (password), something the user has (a smartphone or a hardware token), and something the user is (biometric data such as fingerprints or facial recognition). Users should be educated about the different MFA options available and how to set them up.

  - **Implementation**: Detailed step-by-step guides should be provided to help users enable MFA on their accounts. This may involve configuring MFA for email accounts, social media, and other platforms.

  - **Promoting MFA**: Organizations can encourage MFA usage by making it a standard practice and emphasizing its importance in security training. Additionally, case studies showing the positive impact of MFA in reducing unauthorized access can be included.

Educating users about safe email and web practices not only reduces the risk of delivery-based attacks but also fosters a culture of security within organizations. By arming users with knowledge and skills, we can collectively strengthen our defenses against cyber threats.

Summary

This chapter explored the delivery stage in the cyber kill chain. In this critical phase, cyber adversaries transition from preparation to active engagement by transmitting malicious payloads to their targets. This chapter explained the various delivery methods, including phishing emails, BEC, spear phishing, and drive-by downloads. Real-world incidents, such as the SolarWinds supply chain attack and the Colonial Pipeline ransomware attack, were dissected to illustrate the impact and methodologies of delivery-based cyberattacks. Through these case studies, you gained insight into cybercriminals' sophisticated techniques to bypass security measures and infiltrate target environments.

Furthermore, the chapter highlighted the importance of robust cybersecurity measures and the role of user awareness in defending against these threats. It discussed effective strategies for detecting and blocking delivery attacks, emphasizing the role of web security solutions such as URL filtering, WAFs, and sandboxing. The chapter underscored the necessity of educating users about safe email and web practices, demonstrating how user awareness and training initiatives can serve as a frontline defense. By understanding and implementing these defensive strategies, organizations can significantly mitigate the risks associated with the delivery stage of the cyber kill chain, thereby enhancing their overall cybersecurity posture.

The next chapter will delve into the *Exploitation* stage of the cyber kill chain as we move forward. This phase follows the successful delivery of a malicious payload, where the attacker seeks to exploit vulnerabilities within the target system to gain unauthorized access or execute malicious code. Understanding the intricacies of exploitation techniques and the countermeasures that can be employed to defend against them will be crucial in further fortifying an organization's cybersecurity defenses.

Further reading

Following are the references we used in this chapter along with some bonus resources you can check out:

- `https://blog.virustotal.com/2023/07/virustotal-malware-trends-report.html`
- `https://stonefly.com/blog/how-is-ransomware-delivered/`
- `https://inquest.net/blog/top-malware-delivery-tactics-watch-out-2023/`
- `https://cyberpedia.reasonlabs.com/EN/payload%20delivery.html`
- `https://heimdalsecurity.com/blog/cyber-kill-chain-model/`
- `https://www.todyl.com/blog/cyberattack-lifecycle-delivery`
- `https://kravensecurity.com/cyber-kill-chain/`

Get This Book's PDF Version and Exclusive Extras

UNLOCK NOW

Scan the QR code (or go to `packtpub.com/unlock`). Search for this book by name, confirm the edition, and then follow the steps on the page.

Note: Keep your invoice handly. Purchase made directly from Packt don't require one.

5

Exploitation

The *Exploitation* stage of the Cyber Kill Chain marks a pivotal moment in a cyberattack, where the attacker leverages vulnerabilities within the target system to gain unauthorized access. During this phase, the malicious payload delivered in the preceding stages is activated, allowing the attacker to execute malicious code on the compromised system. This stage is crucial as it transitions the attack from a preparatory phase to an active breach, enabling the attacker to manipulate, exfiltrate, or destroy data. Effective exploitation typically exploits weaknesses in software, applications, or human factors, often bypassing security measures and leading to significant security breaches if not promptly detected and mitigated.

In this chapter, we will delve into the various aspects of the exploitation stage of the cyber kill chain. We will begin by exploring the vulnerabilities commonly targeted by attackers, including software flaws, misconfigurations, and human errors. Next, we will examine the techniques used to exploit these vulnerabilities, ranging from social engineering to advanced malware deployment. Mitigation and defense strategies will be discussed to highlight how organizations can protect themselves against exploitation, including technological solutions and best practices. We will also investigate the psychological principles that attackers exploit to manipulate their targets, making exploitation more effective.

To illustrate these concepts, we will analyze notable case studies such as the Heartbleed OpenSSL vulnerability and the Capital One data breach of 2019, examining how these incidents occurred and the lessons learned from them. The chapter will also cover defensive tactics designed to mitigate exploitation risks, emphasizing the importance of a multi-layered security approach. Finally, we will address the critical role of user education and awareness in preventing exploitation, underscoring how informed and vigilant users can significantly enhance an organization's security posture.

In this chapter, we will delve into the various aspects of the exploitation stage of the cyber kill chain. We will cover the following topics:

- Historical evolution of exploitation techniques
- Type of exploits targeted by the attackers
- Techniques used to exploit these vulnerabilities

- Mitigation and defense strategies

- Psychological principles exploited by attackers

- Case studies, including the Heartbleed OpenSSL vulnerability and the Capital One data breach of 2019

- Defensive tactics to mitigate exploitation risks

- The role of user education and awareness in preventing exploitation

Historical evolution of exploitation techniques

The landscape of cyber exploitation techniques has evolved significantly over the past few decades, driven by advancements in technology, increasing connectivity, and the continuous arms race between attackers and defenders. Understanding this evolution provides valuable context for current and future cybersecurity challenges:

- **Early days of simple exploits and script kiddies**: In the early days of the internet, exploitation techniques were relatively simple and often relied on well-known vulnerabilities in widely used software. Exploits were typically crafted by individuals with moderate technical skills, commonly referred to as "script kiddies." These early exploits included buffer overflows and basic password-guessing attacks.

- **The rise of worms and viruses**: The late 1990s and early 2000s saw the rise of self-replicating malware such as worms and viruses. Notable examples include the Morris Worm (1988) and the ILOVEYOU virus (2000). These exploits spread rapidly through networks, exploiting vulnerabilities in operating systems and applications to propagate themselves. The widespread damage caused by these attacks highlighted the need for improved security practices.

- **Targeted attacks and zero-day exploits**: As defensive measures improved, attackers began to focus on more targeted attacks. The term *zero-day exploit* emerged, referring to previously unknown vulnerabilities that were exploited before vendors had a chance to develop patches. These sophisticated exploits often targeted specific organizations or individuals, marking a shift toward more strategic and impactful cyberattacks. The discovery and sale of zero-day exploits became a lucrative underground market.

- **Exploit kits and automation**: With the proliferation of exploit kits in the mid-2000s, the barrier to entry for cybercriminals was lowered. Exploit kits are pre-packaged sets of tools that automate the exploitation process, making it easier for less skilled attackers to deploy sophisticated attacks. These kits typically target multiple vulnerabilities, increasing the likelihood of successful exploitation. Popular exploit kits such as Blackhole and Angler facilitated widespread cybercrime, including ransomware and data theft.

- **Advanced persistent threats (APTs)**: In the late 2000s and 2010s, the concept of APTs gained prominence. APTs are characterized by prolonged and targeted cyberattacks, often orchestrated by nation-state actors or highly organized cybercriminal groups. These attacks leverage a combination of zero-day exploits, social engineering, and custom malware to infiltrate and maintain a presence within a target network over an extended period. Notable APT campaigns include Stuxnet (2010) and Operation Aurora (2009).

- **Modern exploitation techniques**: Today, exploitation techniques have become more sophisticated and multifaceted. Attackers employ advanced tactics such as the following:

  - **Fileless malware**: Malware that resides in memory rather than being installed on the hard drive, making it harder to detect.

  - **Supply chain attacks**: Exploiting vulnerabilities in third-party software or hardware to compromise multiple organizations.

  - **Living off the land (LotL) attacks**: Leveraging legitimate tools and processes within the target environment to avoid detection.

  - **Ransomware-as-a-service (RaaS)**: Subscription-based models that allow cybercriminals to deploy ransomware attacks with minimal technical knowledge.

The historical evolution of exploitation techniques underscores the dynamic nature of the cybersecurity landscape. As attackers continue to innovate and adapt, defenders must remain vigilant and proactive in their efforts to identify and mitigate emerging threats. Understanding past trends and techniques is essential for developing effective security strategies and staying ahead of potential adversaries.

Type of exploits

In cybersecurity, understanding the different types of exploits is crucial for defense and proactive threat management. Exploits are methods attackers use to take advantage of software, hardware, or network systems vulnerabilities, leading to unauthorized access, data breaches, or system disruptions. Each exploit leverages specific weaknesses, ranging from code execution and privilege escalation to sophisticated social engineering techniques. By recognizing and categorizing these exploits, security professionals can better anticipate potential threats and develop targeted strategies to protect critical systems and data:

- **Remote code execution (RCE)**: RCE exploits allow attackers to run arbitrary code on a remote system, leading to potential complete system compromise. To prevent such exploits, it's crucial to regularly update and patch all software to eliminate known vulnerabilities that can be targeted to execute malicious code on a remote server.

- **Local privilege escalation**: These exploits exploit vulnerabilities to gain higher access levels on a system where the attacker already has limited access. An example includes exploiting an operating system vulnerability to escalate from a standard user to administrative privileges.

- **Buffer overflow**: Buffer overflow exploits occur when an attacker sends more data than a buffer can handle, causing overflow into adjacent memory spaces. This can lead to arbitrary code execution or system crashes. Software applications with inadequate input validation are common targets. The *CrowdStrike Global Threat Report 2024* highlights that buffer overflow remains a critical vulnerability threat actors exploit to gain unauthorized access to systems. As a security professional, IT administrator, or decision-maker, your role is crucial in mitigating this persistent threat vector. Adversaries often target buffer overflow vulnerabilities within unmanaged network appliances, legacy systems, and outdated applications, particularly those lacking modern security measures such as **endpoint detection and response** (**EDR**) sensors. This exploitation allows attackers to execute arbitrary code, escalate privileges, or pivot within the network, often using sophisticated tactics to bypass traditional security controls. Therefore, maintaining up-to-date patches and monitoring for buffer overflow attacks is crucial.

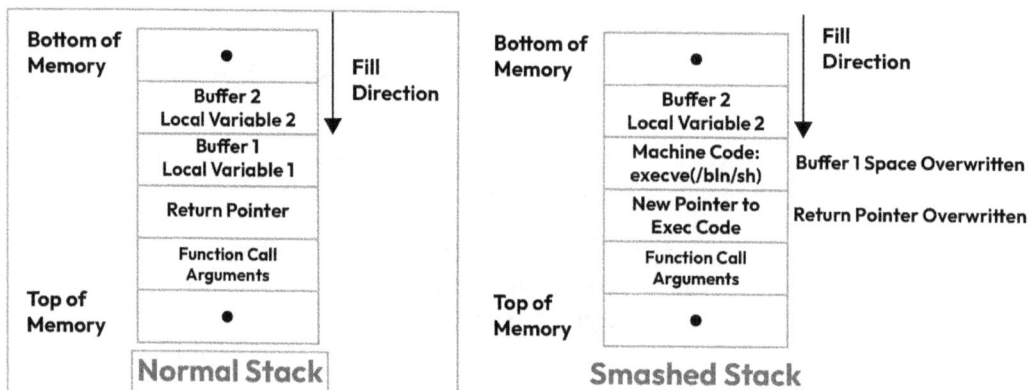

Figure 5.1 – Buffer overflow

- **SQL injection**: SQL injection exploits involve inserting malicious SQL queries into input fields to manipulate databases. This can result in data theft, manipulation, or gaining administrative control over the database, such as exploiting web application vulnerabilities to retrieve sensitive information.

- **Cross-site scripting** (**XSS**): XSS exploits allow attackers to inject malicious scripts into web pages viewed by other users, which can steal session cookies, redirect users to malicious sites, or perform other harmful actions. An example is injecting JavaScript into a web application's input fields that execute in other users' browsers.

- **Cross-site request forgery** (**CSRF**): CSRF exploits trick users into performing actions they did not intend to do on a web application where they are authenticated. For instance, a web application could be exploited to make an authenticated user unknowingly transfer funds or change account settings.

- **Denial of service (DoS) and distributed denial of service (DDoS)**: DoS and DDoS exploits aim to make a system or network resource unavailable to its intended users by overwhelming it with illegitimate requests. This can exploit network protocol or application software vulnerabilities to cause system crashes or resource exhaustion.

- **Man-in-the-middle (MitM)**: MitM exploits occur when an attacker intercepts and potentially alters communication between two parties without their knowledge. For example, vulnerabilities in Wi-Fi networks can be exploited to intercept and modify communications between a user and a website.

MITM Attack

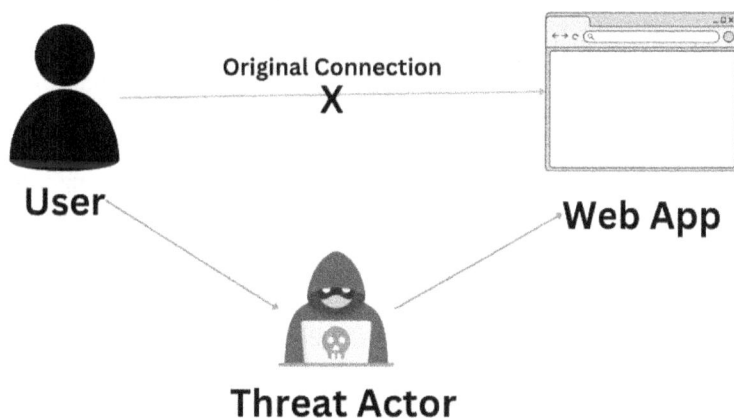

Figure 5.2 – MitM attack

- **Zero-day exploits**: Zero-day exploits target vulnerabilities unknown to the software vendor and for which no patch is available, making them particularly dangerous. They can exploit previously unknown vulnerabilities in widely used software applications before a fix is released.

- **Social engineering exploits**: These exploits rely on manipulating individuals into divulging confidential information or performing actions that compromise security, often complementing technical exploits. Examples include phishing attacks that trick users into revealing passwords or clicking on malicious links.

Understanding these various types of exploits is essential for developing comprehensive security measures. This allows organizations to implement targeted defenses to protect their systems and data from these threats.

Exploitation techniques

The exploitation phase within the Cyber Kill Chain model is a critical stage where cyber adversaries capitalize on identified vulnerabilities to gain unauthorized access to target systems. This phase represents a pivotal moment in the attack life cycle, with attackers strategically leveraging various tactics and techniques. To delve deeper into this phase, we will explore the exploitation of vulnerabilities, the role of zero-day exploits, and different attack vectors. Let's take a look at the techniques:

- **Social engineering**: Social engineering exploits human vulnerabilities rather than technical flaws and is relevant in both the exploitation and installation phases of the cyber kill chain. Attackers use manipulation and psychological tactics to trick individuals into divulging confidential information or performing actions that compromise security. Common social engineering techniques include the following:

  - **Phishing**: Sending fraudulent emails to trick recipients into revealing sensitive information.

  - **Pretexting**: Creating a fabricated scenario to obtain private information.

  - **Baiting**: Offering something enticing to lure victims into providing information or downloading malware.

- **Remote code execution**: RCE is a technique that allows attackers to run arbitrary code on a target system from a remote location. This often leads to full system compromise. Attackers exploit software vulnerabilities to inject and execute malicious code. Key methods include the following:

  - **Buffer overflow**: Overloading a buffer to overwrite adjacent memory and execute arbitrary code.

  - **Code injection**: Inserting malicious code into a vulnerable program to alter its execution.

- **Privilege escalation**: Privilege escalation is a process where attackers gain higher-level access within a system after initial access. This is crucial for expanding their control and carrying out further attacks. The techniques include the following:

  - **Vertical escalation**: Gaining higher privileges (e.g., from user to administrator).

  - **Horizontal escalation**: Accessing other users' accounts with similar privileges.

- **Payload delivery**: Payload delivery involves successfully delivering malicious software to the target system. This step is critical for the execution of the attack. Methods include the following:

  - **Email attachments**: Sending malicious attachments in phishing emails.

  - **Drive-by downloads**: Automatically downloading malware when a user visits a compromised website.

  - **Malware-laden USBs**: Using infected USB drives to spread malware.

- **Zero-day exploits**: They are a grave threat, targeting vulnerabilities that are unknown to the software or hardware vendor and have no available patch. These exploits are particularly dangerous as they can bypass existing defenses. Attackers often use these to launch stealthy and undetectable attacks until the vulnerability is discovered and patched.

- **Misconfiguration exploitation**: Misconfiguration exploitation takes advantage of improperly configured systems. Common targets include the following:

  - **Default passwords**: Exploiting systems that still use factory-set passwords.

  - **Open ports**: Attacking systems with unnecessary open ports.

  - **Unpatched systems**: Targeting systems lacking the latest security updates.

- **Exploiting software flaws**: Attackers often exploit known software flaws that haven't been patched. This includes the following:

  - **SQL injection**: Injecting malicious SQL queries to manipulate databases.

  - **XSS**: Injecting malicious scripts into web pages viewed by other users.

- **Network exploitation**: Network exploitation involves compromising network infrastructure to intercept or manipulate data. Techniques include the following:

  - **MitM attacks**: Intercepting and altering communication between two parties.

  - **DNS spoofing**: Redirecting traffic from legitimate sites to malicious ones.

In the next section, we will focus on the software and system vulnerabilities.

Understanding software and system vulnerabilities

Cybercriminals employ a wide array of techniques to exploit software and system vulnerabilities, making it crucial for cybersecurity professionals to understand these methods to effectively defend against attacks. The *Exploitation* phase of cyberattacks involves the critical step of exploiting vulnerabilities in software and systems. Cybercriminals employ various techniques and tactics to exploit these vulnerabilities, which can have devastating consequences if not adequately defended against.

Let's deep dive to understand software and system vulnerabilities.

What are vulnerabilities?

Vulnerabilities are fundamental aspects of cybersecurity that demand our attention. They are the chinks in the armor of our software, hardware, and configurations that can be exploited by malicious actors to compromise the integrity, confidentiality, or availability of our systems. These vulnerabilities can be found at every layer of a computing environment, from the intricate code of an operating system to the settings in a network firewall. The importance of recognizing vulnerabilities cannot be overstated.

A poignant example is the Equifax breach in 2017, which reverberated throughout the world. This incident was rooted in a vulnerability in Apache Struts, an open source framework for building web applications. The breach affected approximately 143 million individuals and demonstrated the devastating impact that a single vulnerability can have when left unaddressed.

To understand the broader context, vulnerabilities can be classified into several categories. Some of the most common types include design flaws, implementation errors, and system misconfigurations. Design flaws can encompass weaknesses in the architectural decisions of software or hardware, while implementation errors refer to coding mistakes that create unintended entry points for exploitation. System misconfigurations result from errors in setting up or managing systems, often allowing attackers to gain unauthorized access. To secure our digital world, we must delve into the intricate details of these vulnerabilities and learn how to recognize and mitigate them effectively.

Common vulnerabilities

Common vulnerabilities form the bedrock of the cybersecurity landscape, and comprehending these weaknesses is essential for devising robust defense strategies. The tapestry of vulnerabilities is vast, and the threads of insecurity are woven into software, hardware, and configurations across diverse systems. Among the most notable vulnerabilities are buffer overflows, injection attacks, and misconfigured security settings. Let's look into them:

- **Buffer overflows**, a venerable adversary of secure software, occur when an attacker sends more data to a software buffer than it can safely accommodate. This excess data can overwrite critical memory locations, leading to unintended and potentially malicious code execution. The aftermath of buffer overflow exploits has historically wreaked havoc, making it a prime example of a common vulnerability.

- **Injection attacks**, such as SQL injection, are another well-worn path for cybercriminals. These attacks manipulate input fields in web applications, causing the system to execute unintended database queries. Through this, attackers can gain unauthorized access to sensitive information or disrupt the application's normal functioning. Recognizing and mitigating these injection vulnerabilities is pivotal for secure software development and system maintenance.

- **Misconfigured security settings**, while less overt, can be equally devastating. Such misconfigurations may involve exposing critical services to the internet without proper authentication or inadvertently providing overly permissive access to resources. These settings are like unlocked doors in the digital realm, inviting intruders to breach the perimeter. Understanding and rectifying misconfigurations is essential for ensuring the integrity of system defense.

In the next section, we will discuss the common techniques used by cybercriminals, providing insights into their strategies and how organizations can protect themselves against these threats.

Common techniques used by cybercriminals

Cybercriminals continuously adapt and innovate their methods to infiltrate systems, steal sensitive information, and disrupt operations. Understanding the various techniques these malicious actors use is crucial for developing effective defense strategies. This overview delves into cybercriminals' sophisticated tactics and tools, highlighting the importance of awareness and preparedness in the ever-evolving landscape of cybersecurity threats. From social engineering to advanced malware delivery, these techniques underscore the complexity and persistence of cyber threats individuals and organizations face.

Exploiting known vulnerabilities

Cybercriminals are keenly aware of a significant weakness in the cybersecurity landscape – the delay in patching known vulnerabilities. Even when security patches or updates are available, organizations often fail to apply them promptly. This oversight creates an opportunity for attackers to exploit these known vulnerabilities and wreak havoc.

An illustrative case is the WannaCry ransomware attack in 2017. It took advantage of a vulnerability in Microsoft Windows that was addressed by a patch two months prior. However, many organizations hadn't installed the patch, leaving them vulnerable. WannaCry spread rapidly, infecting hundreds of thousands of computers worldwide, including critical systems such as those in the UK's **National Health Service (NHS)**. The attack's scale and impact served as a stark reminder of the consequences of not promptly addressing known vulnerabilities.

The key takeaway is that cybercriminals actively monitor the release of security patches and target organizations with delayed patching processes. To mitigate this risk, it's essential for organizations to establish rigorous patch management procedures and ensure timely application of patches to minimize their exposure to exploitation.

Zero-day exploits

Zero-day exploits are among the most potent weapons in the arsenal of cybercriminals. These exploits target vulnerabilities that are unknown to the software vendor or the public. The name "zero-day" refers to the fact that on the day the vulnerability is discovered, there are zero days during which the vendor can release a patch to fix it.

A standout example of a zero-day exploit is the Stuxnet worm, discovered in 2010. Stuxnet was a highly sophisticated piece of malware, primarily designed for cyber espionage and the disruption of Iran's nuclear program. It exploited multiple zero-day vulnerabilities in Windows operating systems and industrial control systems. Because these vulnerabilities were not publicly known, there were no security patches available to defend against the worm. The successful use of zero-day exploits in Stuxnet underscores their potential for causing significant damage.

Zero-day exploits are particularly challenging to defend against since there is no advance notice or available patch. To counter these threats, organizations must focus on advanced threat detection and response, closely monitoring network behavior for signs of suspicious activity.

One of the earliest instances of a buffer overflow attack is associated with the Morris Worm in 1988. The Morris Worm exploited vulnerabilities in the Unix fingerd and sendmail programs, using buffer overflows to propagate itself across the early internet. The worm's unintended impact was to highlight the significance of buffer overflow vulnerabilities, prompting the development of countermeasures such as stack canaries and **address space layout randomization (ASLR)**.

Understanding the mechanics of buffer overflow attacks is crucial for defense. Today, secure coding practices, regular code reviews, and tools such as static code analyzers are employed to detect and prevent buffer overflow vulnerabilities in software.

SQL injection

SQL injection is a prevalent technique employed by cybercriminals to manipulate database queries through input fields, ultimately gaining unauthorized access or extracting sensitive information. A high-profile example of a successful SQL injection attack is the Ashley Madison data breach in 2015.

In this breach, attackers exploited a vulnerability in the Ashley Madison website, which allowed them to inject malicious SQL code into the login form. This breach exposed sensitive user data, leading to public embarrassment and potential legal consequences for the company. Recognizing and mitigating SQL injection attacks is of paramount importance, particularly for websites and applications that handle user inputs.

Defense against SQL injection involves rigorous input validation, employing prepared statements or parameterized queries, and the use of **web application firewalls (WAFs)** designed to detect and block such attacks.

Credential reuse

Credential reuse occurs when individuals use the same username and password combinations across multiple websites or services, posing significant security risks. Cybercriminals exploit this practice through credential stuffing, where automated tools test stolen credentials from one breach across multiple sites, leading to unauthorized access and financial losses. This increases vulnerability, as a single data breach can cascade and compromise various platforms, amplifying the damage. Consequently, the widespread use of reused credentials weakens security across all affected systems, making it easier for cybercriminals to infiltrate and exploit personal and organizational data.

Brute force

The **brute force attack** methodology lies at the extreme end of the exploitation spectrum. This approach entails the systematic attempt of all conceivable permutations to gain unauthorized access to a system or decrypt protected data. While potentially efficacious, this method is frequently accorded lower priority in a malicious actor's strategic framework, primarily due to two significant factors. Firstly, its time-consuming and resource-intensive nature renders it less efficient than exploiting known vulnerabilities. Secondly, there is an elevated risk of detection stemming from potential security alerts triggered by repeated unsuccessful attempts. Consequently, malicious actors prioritize more discreet and efficient attack vectors before resorting to brute force techniques.

Exploitation framework

Imagine a world where hackers and security professionals are constantly battling wits, each armed with a unique set of tools. **Exploitation frameworks and tools** are the weapons of choice in this digital battlefield, enabling ethical hackers and malicious attackers to discover, develop, and execute exploits to gain unauthorized access to systems or data. Some of the most notable frameworks include the following:

- **Metasploit**: A comprehensive library of exploits, payloads, and auxiliary modules

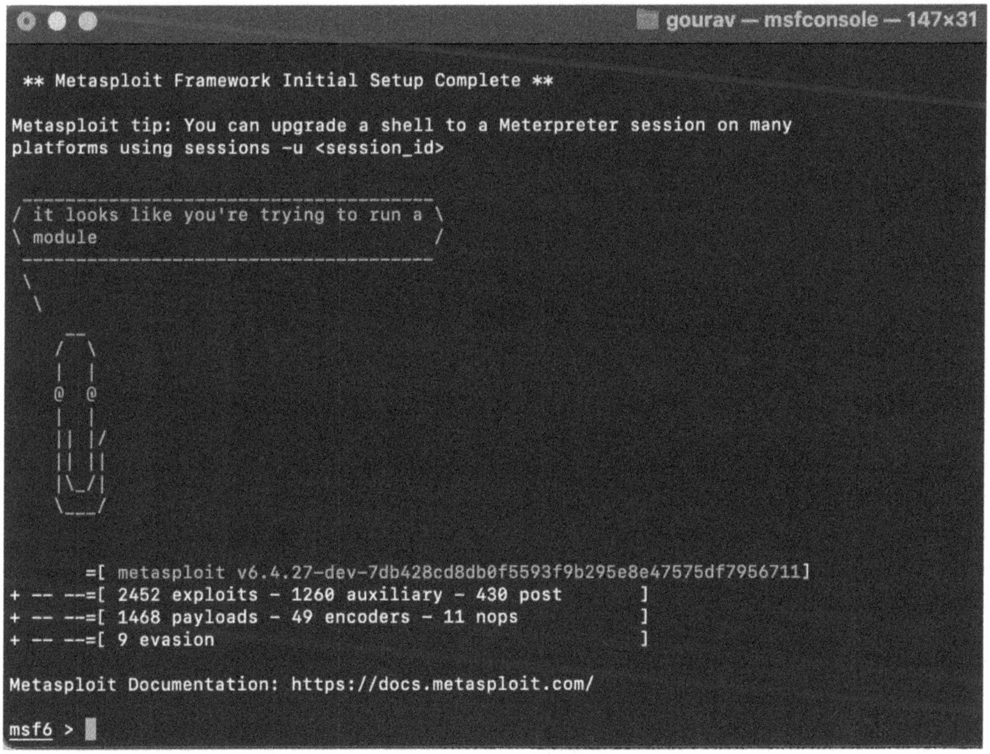

Figure 5.3 – Metasploit framework

- **Core Impact**: A commercial penetration testing platform offering automated and manual testing capabilities

- **CANVAS by Immunity Inc.**: A flexible platform for creating custom attacks with extensive pre-built exploits

- **Cobalt Strike**: Used in red team operations, simulating APTs

- **Empire**: Leveraging PowerShell and Python for post-exploitation activities

- **SQLMap**: An open source tool for automating the detection and exploitation of SQL injection vulnerabilities

- **Burp Suite**: A comprehensive tool for web application security testing

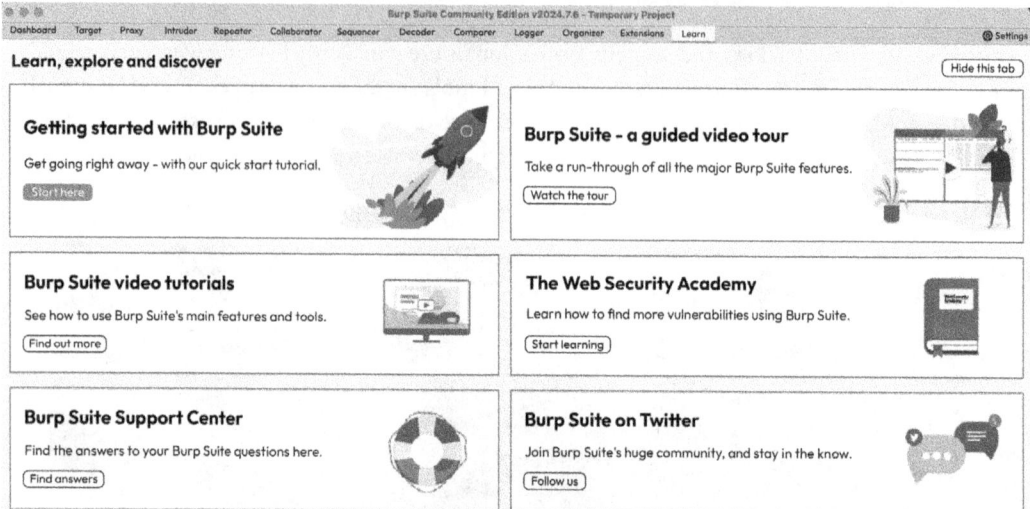

Figure 5.4 – Burp Suite homepage

Psychological principles exploited by attackers

Cyber attackers often rely on psychological manipulation to bypass technical defenses and exploit human vulnerabilities. By understanding and leveraging psychological principles, attackers can craft highly effective social engineering attacks that deceive individuals into divulging sensitive information or performing actions that compromise security. Let's explore some of the fundamental psychological principles exploited by attackers:

- **Authority**: The principle of authority involves exploiting people's tendency to comply with figures of authority. Attackers often impersonate authority figures, such as executives, IT personnel, or government officials, to gain trust and prompt immediate action. For example, an attacker might send an email pretending to be a CEO, instructing an employee to transfer funds urgently or disclose confidential information.

- **Reciprocity**: Reciprocity is the human inclination to return favors or kindnesses. Attackers exploit this principle by offering something of value, such as a gift, exclusive information, or assistance, to induce the target to reciprocate by sharing sensitive information or performing a requested action. Phishing emails often offer free downloads or access to premium content in exchange for login credentials.

- **Scarcity**: Scarcity taps into the **fear of missing out (FOMO)** by creating a sense of urgency or limited availability. Attackers use this principle to pressure targets into making quick decisions without proper scrutiny. For example, a phishing email might claim that an account will be suspended unless immediate action is taken, prompting the recipient to click on a malicious link or provide personal information.

- **Social proof**: Social proof involves leveraging individuals' tendency to follow others' actions, especially in uncertain situations. Attackers may use fake testimonials and endorsements or mimic common behaviors to build credibility and influence targets. For instance, an email may come from a trusted colleague or be designed to look like a popular service, encouraging recipients to follow instructions without question.

- **Commitment and consistency**: Once people commit to something, they are more likely to follow through and remain consistent with their previous actions or beliefs. Attackers exploit this principle by starting with small requests that seem harmless and gradually escalating to more significant requests. This technique is often seen in phishing attacks, where initial benign interactions lead to increasingly sensitive requests.

- **Liking**: People are more likely to comply with requests from individuals they like or feel connected to. Attackers build rapport and establish a sense of familiarity or likability to manipulate targets. This can be achieved through personalized messages, friendly language, or pretending to share common interests. Social media platforms frequently gather information about targets to craft convincing and likable personas.

- **Fear and urgency**: Fear and urgency are powerful motivators that can override rational decision-making. Attackers exploit these emotions by creating scenarios that invoke fear, such as threats of account closure, legal action, or security breaches. The urgent need to address the perceived threat prompts targets to act quickly, often leading to the disclosure of sensitive information or the execution of harmful actions.

Case studies – impactful exploitation incidents

In this section, we will examine notable case studies that illustrate the real-world impact of exploitation incidents. These case studies provide valuable insights into the methods used by attackers, the vulnerabilities exploited, and the lessons learned. By understanding these incidents, you can better appreciate the importance of robust security measures and proactive defense strategies, which can help prevent similar incidents in the future:

- **Case study 1 – Heartbleed OpenSSL vulnerability (2014)**: The Heartbleed vulnerability was a critical flaw in the OpenSSL cryptographic library, allowing attackers to access sensitive data from affected servers. We will analyze its origins, impacts, and the lessons it imparted to the cybersecurity community.

- **Case study 2 – The Capital One data breach (2019)**: This exposed sensitive customer information due to a misconfigured WAF. We will explore the execution of the breach, its impact, and the critical lessons for cloud security.

Case study 1 – Heartbleed OpenSSL vulnerability (2014)

The Heartbleed OpenSSL vulnerability, discovered in April 2014, stands as a watershed moment in the history of cybersecurity. This section will provide a comprehensive analysis of the Heartbleed vulnerability, examining its origins, impacts, and the lessons it imparted to the cybersecurity community.

The Heartbleed vulnerability was not just another software flaw but a critical vulnerability in the widely used OpenSSL cryptographic library, specifically in its implementation of the **Transport Layer Security (TLS)** and **Secure Sockets Layer (SSL)** protocols. OpenSSL is fundamental for secure communication on the internet, ensuring the confidentiality and integrity of data transmitted over networks.

Heartbleed exploited a programming error in the *OpenSSL Heartbeat Extension*, hence its name. The bug allowed an attacker to send a malicious heartbeat request to a vulnerable server, tricking it into revealing random segments of its memory, which often contained sensitive data. The vast scope of OpenSSL's usage made the Heartbleed vulnerability particularly concerning. It affected millions of web servers worldwide, including popular websites and services, such as Yahoo, GitHub, and Canadian tax services. Even network devices, email servers, and VPNs were at risk.

The consequences of Heartbleed were profound, demonstrating how a single vulnerability could ripple across the internet, impacting organizations and individuals alike.

The Heartbleed bug allowed attackers to access sensitive data, including usernames, passwords, encryption keys, and more, from affected servers. The potential for data breaches and identity theft was significant.

In many cases, organizations were forced to reissue security certificates and change encryption keys, a time-consuming and costly process. This was essential to prevent attackers from eavesdropping on future communications.

Heartbleed exposed the challenges associated with patch management. While a fix was made available promptly, many organizations struggled to apply it swiftly due to the sheer number of affected servers.

The Heartbleed incident left a lasting impact on cybersecurity practices and policy. Heartbleed highlighted the critical role of open source software in the digital ecosystem. It also emphasized the need for transparent security practices and better resourcing for open source projects.

The incident underscored the significance of robust vulnerability management. Organizations began to focus on proactive vulnerability scanning, patch management, and incident response planning.

User education about secure online practices became a priority. Heartbleed served as a reminder of the importance of strong, unique passwords and multi-factor authentication.

The Heartbleed OpenSSL vulnerability of 2014 was a wake-up call for the cybersecurity community, demonstrating the far-reaching impact a single vulnerability can have. It prompted changes in open source security, vulnerability management, and user awareness, providing valuable lessons for a more secure digital future.

Case study 2 – The Capital One data breach (2019)

In July 2019, Capital One, a leading financial institution, fell victim to a data breach that sent shockwaves through the cybersecurity landscape. This incident underscored the vulnerabilities that organizations face in the age of cloud computing and emphasized the critical need for robust cloud security configurations and effective vendor risk management.

The Capital One data breach was a massive cybersecurity incident that affected approximately 106 million customers in the United States and Canada. At the center of this breach was the exposure of sensitive customer information, including names, addresses, credit scores, social security numbers, and bank account numbers. The breach had significant repercussions, both for Capital One and the broader financial industry.

The breach was executed by a former employee of a cloud services provider, specifically, **Amazon Web Services (AWS)**. The attacker, a former AWS employee who had insider knowledge, took advantage of a misconfigured WAF to gain unauthorized access to Capital One's customer data. The misconfiguration in the WAF allowed the attacker to exploit a vulnerability in the system and penetrate Capital One's cloud infrastructure hosted on AWS.

The attacker leveraged this access to exfiltrate sensitive customer data. This breach highlighted the potential risks associated with insider threats, as a former employee with knowledge of the cloud infrastructure's intricacies was able to exploit vulnerabilities effectively.

The impact of the Capital One data breach was substantial, affecting millions of customers and casting a shadow of doubt over the security of cloud-based systems. The breach exposed sensitive personal and financial information of customers, which could lead to identity theft, financial fraud, and other malicious activities.

In response to the breach, Capital One had to allocate significant resources to mitigate the damage, investigate the extent of the incident, and enhance its security measures. Furthermore, the breach incurred substantial financial costs, including legal settlements and regulatory penalties.

The Capital One data breach serves as a significant case study with several critical lessons for organizations.

Cloud security is a shared responsibility between cloud service providers and their customers. While cloud providers such as AWS offer robust security features, customers must configure and manage their cloud environments properly. The misconfiguration of the WAF was a pivotal factor in this breach, emphasizing the need for thorough security assessments and best practices.

The breach highlighted the real and significant threat posed by insider threats. Even a former employee with knowledge of the cloud infrastructure's configuration can exploit vulnerabilities. This necessitates organizations to implement strict access controls, conduct regular reviews of permissions, and monitor employee activities.

Organizations need to assess and manage the security of third-party vendors and service providers who have access to their data. The breach was facilitated by the compromise of a cloud services provider's infrastructure, highlighting the need for rigorous vendor risk assessments and security due diligence.

A robust incident response plan is essential for minimizing the impact of data breaches. Capital One's response to the breach included swift action to contain the incident, notify affected customers, and cooperate with law enforcement agencies. Timely and effective incident response is critical in such situations.

Capital One faced regulatory investigations and legal actions due to the breach. Organizations must be prepared to navigate the legal and regulatory landscape in the aftermath of a breach.

The Capital One data breach of 2019 remains a cautionary tale that highlights the evolving threat landscape in cloud-based systems. It serves as a reminder of the critical importance of proper cloud security configurations, rigorous vendor risk management, and proactive security measures to protect customer data and maintain trust in the digital age.

Mitigation and defense strategies

Understanding and mitigating the techniques used by cybercriminals is crucial for maintaining robust cybersecurity defenses. Organizations must stay vigilant and proactive as attackers continually evolve their methods to exploit vulnerabilities. This overview delves into essential defense strategies, including regular patching, vulnerability scanning, WAFs, intrusion detection systems, and user training. By implementing these measures, along with additional techniques such as network segmentation, EDR, multi-factor authentication, **security information and event management** (**SIEM**), and regular security audits, organizations can significantly enhance their resilience against cyber threats, reduce their attack surface, and safeguard their systems and data from unauthorized access and malicious activities:

- **Regular patching**: The timely application of security patches and updates is fundamental for defending against known vulnerabilities. Software vendors release patches in response to identified security flaws, and failing to apply them promptly can expose systems to exploitation. Attackers often target these known vulnerabilities. The U.S. Department of Homeland Security emphasizes the importance of a robust patch management process to minimize security breaches. This involves regularly monitoring for new patches, testing them in a controlled environment to ensure they don't disrupt operations, and applying them promptly. Organizations can significantly reduce their attack surface and enhance their resilience against cyber threats by keeping software and systems updated with the latest security patches.

- **Vulnerability scanning and assessment**: Regular vulnerability assessments and scanning are crucial for identifying weaknesses in systems and software. Tools such as Nessus and OpenVAS systematically scan networks and systems for known vulnerabilities, misconfigurations, and security weaknesses, providing detailed reports highlighting the vulnerabilities and their severity. These results enable organizations to prioritize their efforts by first addressing the most critical vulnerabilities. Vulnerability scanning is an ongoing process due to the continuously evolving threat landscape. Periodic assessments help maintain a strong security posture. Additionally, organizations may consider penetration testing, where ethical hackers simulate real-world attacks to assess system security comprehensively.

- **WAFs**: These protect web applications from common vulnerabilities, including SQL injection and XSS attacks. WAFs use signature-based analysis to compare incoming requests to known attack patterns and behavior-based analysis to identify anomalies in application behavior that may indicate an attack. Deploying a WAF adds additional protection to web applications, filtering out malicious traffic and safeguarding application data. Regular updates and rule configuration are essential to keep WAFs effective against emerging threats. Another preventative control is instruction sanitization, which is the process of cleaning and validating input data to remove or neutralize potentially malicious code or instructions. This technique helps prevent attacks such as SQL injection and XSS by filtering out dangerous elements before they reach the web application.

- **Intrusion detection systems (IDSs)**: IDSs are vital for identifying and mitigating exploitation attempts. IDSs monitor network traffic for suspicious activity and known attack patterns, alerting security teams to potential threats. For a deeper understanding of IDSs, organizations can refer to the *Intrusion Detection Systems Guide* provided by the **U.S. Computer Emergency Readiness Team (US-CERT)**.

- **Intrusion prevention systems (IPSs)**: IPSs are network security tools that monitor, detect, and actively prevent threats. Unlike an IDS, which only alerts, an IPS can automatically block or prevent malicious activities that are detected in real time.

- **Security awareness training**: Educating users about safe practices and recognizing social engineering attempts is crucial for defense against exploitation. Effective user training programs can significantly reduce the risk of human error leading to security breaches. The *Security Awareness and Training* section of the NIST Cybersecurity Framework provides guidelines for developing such programs.

  Security awareness training is an indispensable component of safeguarding against social engineering attacks. It equips employees with the knowledge and skills to recognize, resist, and respond to manipulative tactics employed by attackers. Here, we will explore the nuances of this training, emphasizing its significance in enhancing cyber resilience. An essential aspect of effective security awareness training is interactive simulations. These simulations replicate real-world scenarios where employees encounter social engineering attempts. By actively engaging with these simulations, individuals can experience the psychological pressure and decision-making process involved in such situations. This experiential learning approach deepens their understanding of the risks and equips them with practical skills. Following are the most important aspects to cover in security awareness training:

  - **Zero Trust framework**: Implementing a Zero-Trust security model can significantly enhance defenses during this critical phase. Zero Trust is a security framework that operates on the principle of "never trust, always verify," meaning that no entity — internal or external — is trusted by default. Here, we explore how Zero-Trust principles can be applied to defend against exploitation.

- **Network segmentation**: Network segmentation involves dividing a network into smaller, isolated segments to limit the spread of malware and restrict unauthorized access. By segmenting critical systems and sensitive data, organizations can reduce the impact of a breach and prevent attackers from moving laterally across the network. This technique also helps apply more granular security policies to different segments based on risk levels.

- **Multi-factor authentication (MFA)**: MFA adds an extra layer of security by requiring users to provide two or more verification factors to gain access to resources. This significantly reduces the risk of unauthorized access due to compromised credentials. MFA can include a combination of something the user knows (password), something the user has (security token), and something the user is (biometric verification).

- **Regular security audits and penetration testing**: Regular security audits and penetration testing involve systematically evaluating an organization's security measures and simulated attacks to identify and address vulnerabilities. Security audits assess compliance with security policies and standards, while penetration testing identifies exploitable weaknesses through controlled attacks. Conducting these activities regularly helps maintain a strong security posture and ensures continuous improvement in defense mechanisms.

Defending against cyber threats requires a multi-layered approach addressing technical vulnerabilities and human factors. By implementing a combination of regular patching, vulnerability scanning, WAFs, IDSs, and comprehensive user training, organizations can build a robust defense against cybercriminal activities. Additionally, advanced techniques such as network segmentation, EDR, MFA, SIEM, and regular security audits further enhance security resilience. Understanding and mitigating the exploitation techniques cybercriminals use is essential for protecting systems and data, ensuring a secure and resilient cybersecurity posture.

Ethical considerations and legal implications

Understanding the ethical considerations and legal implications is crucial in cybersecurity, especially when dealing with exploitation techniques and vulnerabilities. These aspects ensure that security practices are conducted responsibly and within the bounds of the law, thereby protecting both the professionals involved and the organizations they work with.

When discussing the ethical considerations in cybersecurity, two practices stand out: responsible disclosure and bug bounty programs. These practices ensure that security vulnerabilities are identified and addressed responsibly:

- **Responsible disclosure**: Responsible disclosure is a practice where security researchers and ethical hackers privately report discovered vulnerabilities to the affected organization or vendor, allowing them to address the issue before it is publicly disclosed. This approach balances the need to inform the public about security risks and protect systems from malicious exploitation:

  - **Ethical considerations**: Ensures that malicious actors do not exploit vulnerabilities and that organizations can patch vulnerabilities before they become public knowledge.

- **Legal implications**: This policy protects researchers from potential legal repercussions for publicly disclosing vulnerabilities without allowing the affected party to fix them. It also helps organizations comply with data protection laws and regulations by addressing security flaws promptly.

- **Bug bounty programs**: Bug bounty programs are initiatives where organizations reward security researchers for identifying and reporting security vulnerabilities in their software or systems. Some of the popular bug bounty vendors are HackerOne and BugCrowd. These programs incentivize ethical hacking and help organizations discover and mitigate vulnerabilities:

  - **Ethical considerations**: Participants in bug bounty programs should adhere to the program's rules and guidelines, avoiding actions that could damage or disrupt the system.

  - **Legal implications**: Organizations must clearly define the scope and rules of their bug bounty programs to prevent legal issues. Participants should ensure that their actions are within the legal boundaries set by the program.

These practices and considerations help protect sensitive data and systems and foster a collaborative environment where vulnerabilities can be safely identified and addressed. Stay tuned; in future chapters, we will do a deep dive into the ethical considerations of cybersecurity.

Future trends in exploitation techniques

As technology evolves, so do the methods and techniques used by cybercriminals to exploit vulnerabilities. Understanding these future trends is crucial for staying ahead in the cybersecurity landscape. Here are some anticipated future trends in exploitation techniques:

- **Increased use of artificial intelligence and machine learning**: Attackers are increasingly leveraging AI and **machine learning** (**ML**) to automate discovering and exploiting vulnerabilities. Key trends include the following:

  - **Automated vulnerability detection**: AI-driven tools that can rapidly scan and identify vulnerabilities in systems and applications. Automated vulnerability detection tools, such as DeepLocker, are already demonstrating the potential of AI in crafting sophisticated attacks.

  - **Adaptive malware**: Malware can change its behavior based on the environment it finds itself in, making it harder to detect and counter.

  - **Deep learning for phishing**: Using AI to craft more convincing phishing emails by analyzing vast amounts of data to mimic legitimate communications.

- **Exploiting IoT devices**: **Internet of Things** (**IoT**) devices present a growing attack surface due to their widespread adoption and often inadequate security measures. Trends include the following:

  - **Botnets**: Compromising IoT devices to form large botnets for launching DDoS attacks. The Mirai Botnet attack in 2016 showcased the vulnerabilities in IoT devices, leading to massive DDoS attacks.

  - **Home network exploits**: Targeting smart home devices to gain access to home networks and sensitive personal information.

  - **Industrial IoT (IIoT) exploits**: Attacking IIoT devices used in critical infrastructure and industrial systems.

- **APTs**: APTs involve sophisticated, long-term attacks typically sponsored by nation-states or highly organized cybercriminal groups. Future trends include the following:

  - **Supply chain attacks**: Infiltrating the software supply chain to insert malicious code into widely used software products. The SolarWinds attack is a prime example of how supply chain attacks can infiltrate high-value targets.

  - **Stealth techniques**: Enhanced use of evasion techniques to remain undetected within a target network for extended periods.

  - **Targeted ransomware**: Ransomware is specifically designed to target high-value organizations and systems demanding higher ransoms.

- **Quantum computing**: Quantum computing is poised to revolutionize many fields, including cybersecurity. However, it also presents new challenges:

  - **Breaking encryption**: Quantum computers may eventually be able to break current encryption algorithms, necessitating the development of quantum-resistant cryptography. While still emerging, the potential of quantum computing to break traditional encryption was demonstrated by Google's quantum supremacy announcement in 2019. Recently, NIST has released new standards for post-quantum cryptography, and this is covered in the last chapter of this book.

  - **Enhanced cryptographic exploits**: Using quantum computing to develop new types of cryptographic attacks infeasible with classical computers.

- **Cloud exploitations**: As organizations increasingly move to the cloud, attackers are developing new techniques to exploit cloud environments:

  - **Cloud misconfigurations**: Exploiting misconfigured cloud services to gain unauthorized access to data and systems. The Capital One data breach in 2019 highlights the risks associated with cloud misconfigurations.

- **Credential harvesting**: Attacking cloud-based services to steal login credentials, enabling access to multiple interconnected systems.

- **Container exploits**: Targeting vulnerabilities in containerized applications and orchestration platforms like Kubernetes.

- **Social engineering and human factors**: Despite technological advancements, human factors remain a significant vulnerability:

  - **Deepfake technology**: Using AI to create realistic but fake audio and video to impersonate individuals and deceive targets. Deepfake technology was notably used to impersonate a CEO, leading to a fraudulent transfer of $243,000.

  - **Sophisticated phishing campaigns**: More personalized and convincing phishing attacks that leverage social media and publicly available data.

  - **Psychological manipulation**: Advanced techniques to exploit psychological principles more effectively, such as trust, authority, and urgency.

- **Fileless malware**: Fileless malware operates without relying on traditional executable files, making it harder to detect and remove:

  - **Memory-resident malware**: Malware that resides in system memory rather than on disk, evading traditional antivirus solutions. The attack on the Democratic National Committee in 2016 utilized fileless malware, making it difficult to detect and attribute.

  - **Living-off-the-land attacks**: Using legitimate tools and processes already in the system to carry out malicious activities.

- **Enhanced evasion techniques**: Attackers are continually developing new methods to evade detection and forensic analysis:

  - **Polymorphic malware**: Malware that changes its code with each infection to avoid signature-based detection.

  - **Encrypted communication**: Using encrypted channels to communicate with **command and control (C2)** servers, which makes it difficult for defenders to intercept and analyze.

Rapid technological advancements and evolving threat landscapes shape the future of exploitation techniques. Staying informed about these trends and adapting security measures is essential for defending against increasingly sophisticated cyber threats. By anticipating these developments, organizations can better prepare for the challenges and enhance their overall cybersecurity posture.

Summary

This chapter explored the critical *Exploitation* phase of the Cyber Kill Chain model, where attackers leverage identified vulnerabilities to gain unauthorized access and execute malicious activities. We discussed vital defense strategies, including regular patching, vulnerability scanning, web application firewalls, intrusion detection systems, and user training, which help organizations strengthen their security posture. Additionally, advanced techniques such as network segmentation, endpoint detection and response, multi-factor authentication, security information and event management, and regular security audits were highlighted as essential for enhancing cybersecurity resilience. Understanding these exploitation techniques and implementing robust defense mechanisms are crucial for mitigating cyber threats and protecting organizational systems and data.

Having established a thorough understanding of the *Exploitation* phase, we move on to the *Installation* phase in the Cyber Kill Chain model in the next chapter. In this phase, attackers aim to install malicious software or tools on the compromised system to maintain persistent access and control. We will investigate various installation techniques, including malware, backdoors, and remote access tools. We will explore practical strategies to detect and prevent unauthorized installations, strengthening an organization's security defenses.

Further reading

Following are some bonus reading materials for you to check out:

- *Explains what script kiddies are and their impact on cybersecurity*: https://mergebase.com/cyberbasics/what-is-a-script-kiddie/

- *CrowdStrike's 2024 Global Threat Report highlighting key cybersecurity trends and insights*: https://www.crowdstrike.com/global-threat-report/

- *NIST announces the release of three finalized post-quantum encryption standards*: https://www.nist.gov/news-events/news/2024/08/nist-releases-first-3-finalized-post-quantum-encryption-standards

- *Explains buffer overflow vulnerabilities, their impact, and mitigation strategies*: https://www.cloudflare.com/learning/security/threats/buffer-overflow/

- *Download the page for Metasploit, a widely used penetration testing tool*: https://www.metasploit.com/download

- *PortSwigger's website offering web application security software and learning resources*: https://portswigger.net/

Get This Book's PDF Version and Exclusive Extras

UNLOCK NOW

Scan the QR code (or go to packtpub.com/unlock). Search for this book by name, confirm the edition, and then follow the steps on the page.

Note: Keep your invoice handly. Purchase made directly from Packt don't require one.

6

Installation

The Cyber Kill Chain *Installation* phase represents a pivotal moment in which attackers secure a foothold within a target system by deploying malicious software. Following the initial stages of *Reconnaissance, Weaponization, and Delivery* discussed in previous chapters, the installation phase marks the point where the attack becomes deeply entrenched. This chapter will provide an in-depth overview of this critical phase. It will explore various installation methods, the techniques attackers use to embed malware, and how they ensure persistence within compromised networks. Key topics include persistence mechanisms, tools and technologies employed by attackers and defenders, and essential aspects of incident response. Persistence mechanisms will be examined, showcasing how attackers maintain access over time, including advanced techniques like rootkits and bootkits to evade detection. Additionally, the chapter will investigate the tools and technologies employed by both attackers and defenders, offering insights into the evolving landscape of cyber threats.

Furthermore, this chapter will cover essential aspects of incident response, providing strategies for identifying and addressing installation attempts swiftly and effectively. The topics covered in this chapter are:

- Techiques of installation
- Common tools and technologies used at the installation stage
- Droppers and payload installation
- Defensive strategies for malware
- Notable incidents involving malware installation
- Security measures to prevent unauthorized installations
- Incident response and recovery after malware installation
- The role of artificial intelligence and machine learning in preventing malware installation

Techniques of installation

There are many techniques in the Cyber Kill Chain that attackers use to place malicious software onto a target system. These methods can vary in complexity and sophistication, but they typically fall into the following categories:

- **Phishing emails and malicious attachments**: Attackers use deceptive emails to trick victims into opening attachments or clicking on links that download and install malware. These emails often come from legitimate sources and may contain urgent or enticing messages to prompt quick action.

- **Drive-by downloads**: This method involves compromising a legitimate website to deliver malware to visitors without their knowledge. When a user visits the infected site, the malware is automatically downloaded and installed on their system through vulnerabilities in the web browser or plugins.

- **Exploit kits**: Exploit kits are automated tools that scan for and exploit software vulnerabilities to deliver malware. They are often used with drive-by downloads and can effectively target unpatched systems.

- **Watering hole attacks**: Attackers identify websites frequently visited by their target audience and compromise them to deliver malware. The malware is installed on their system when the target visits the infected site. This method leverages the trust users have in these frequently visited sites.

- **Remote access tools (RATs)**: RATs are designed to allow attackers to remotely control the compromised system. They are often delivered through phishing emails, drive-by downloads, or other methods, allowing attackers to install and manage additional malware on the target system.

- **Supply chain attacks**: Attackers compromise a legitimate software vendor or service provider to insert malware into software updates or applications. When users download and install these seemingly legitimate updates, they unknowingly install the malware.

- **Social engineering**: Attackers manipulate individuals into performing actions that lead to malware installation. This can include tricking users into downloading and installing malicious software by disguising it as a legitimate application or utility.

- **Malvertising**: This technique involves embedding malicious code within online advertisements. When users view or click on these ads, the malware is downloaded and installed on their systems, often exploiting vulnerabilities in the web browser or its plugins.

- **Physical access**: In some cases, attackers may gain physical access to a target system to install malware directly. This method is less common but highly effective, particularly in environments with lax physical security controls.

By understanding these techniques, defenders can better anticipate potential threats and implement effective countermeasures to protect their systems from malicious installations.

Common tools and technologies used at the installation stage of the Cyber Kill Chain

The *Installation* stage of the Cyber Kill Chain is pivotal. In this stage, adversaries solidify their control over a compromised system by deploying malicious software. This phase ensures they maintain persistent access, allowing ongoing exploitation and command execution. The tools and technologies used during this stage are sophisticated and varied, reflecting the evolving landscape of cyber threats. Let's dive into the details:

- **Remote Access Trojans (RATs)**: Remote Access Trojans, commonly known as RATs, are the go-to tools for attackers seeking to control a compromised machine remotely. These tools offer a plethora of functionalities, making them highly effective and dangerous. Let's look into some RATs:

  - **DarkComet**: This RAT has garnered a reputation for its reliability and extensive feature set. It allows attackers to manipulate files, capture screenshots, log keystrokes, and even control the webcam. Its user-friendly interface makes it a favorite among cybercriminals.

  - **NanoCore**: Known for its rich capabilities, NanoCore is often employed for its ability to perform detailed system surveillance, including screen capture, keylogging, and password theft. Its modular design enables attackers to customize its functions according to their needs.

  - **Poison Ivy**: Despite being older, Poison Ivy remains a potent tool. It provides comprehensive control over the infected system, allowing for file manipulation, process management, and remote shell access. Its simplicity and effectiveness keep it in circulation.

- **Backdoors**: Backdoors are crucial for maintaining long-term access to compromised systems. They enable attackers to re-enter the system even after initial detection and cleanup:

  - **Metasploit**: This well-known penetration testing framework includes numerous backdoor capabilities. Attackers can use it to establish persistent access by embedding backdoors into compromised systems, ensuring they can regain entry whenever needed.

  - **Cobalt Strike**: Used mainly for adversary simulations and red teaming, Cobalt Strike also provides robust backdoor functionalities. It helps attackers set up covert channels for data exfiltration and remote command execution.

- **Persistence Mechanisms**: Attackers must maintain persistence to ensure their malicious presence isn't easily eradicated. Various mechanisms, such as the following, are employed to achieve this:

  - **Service Creation**: Installing malware as a service allows it to start automatically with system boot. This method is particularly effective in environments where services are regularly monitored and maintained.

- **Registry Entries**: Attackers often create registry entries that instruct the operating system to execute the malware every time the system starts. This persistence mechanism ensures that the malware remains active, even if the system is rebooted. Research by FireEye has shown that registry-based persistence is a prevalent technique used by **advanced persistent threat (APT)** groups. These findings emphasize the importance of understanding this persistence strategy.

- **Scheduled Tasks**: Another method of achieving persistence is through scheduled tasks. Attackers can create scheduled tasks that regularly execute the malware, allowing them to maintain control over the compromised system. The U.S. **Cybersecurity and Infrastructure Security Agency (CISA)** has documented instances where attackers used scheduled tasks to establish persistence. This highlights the relevance of recognizing scheduled tasks as a persistence strategy in the *Installation* phase.

- **Packers and Crypters**: Packers and Crypters are tools used to obfuscate and encrypt malicious payloads, making them harder to detect and analyze:

  - **Ultimate Packer for Executables (UPX)**: UPX is a popular tool that compresses executables, reducing their size and adding a layer of obfuscation. This makes it more challenging for security software to detect the packed malware.

  - **Veil-Evasion**: This framework is designed to generate payloads that can bypass common antivirus solutions. It employs various techniques to evade detection, such as code obfuscation and encryption.

- **Command and Control (C2) Infrastructure**: Establishing a robust command and control infrastructure is critical for attackers to communicate with compromised systems and orchestrate further actions:

  - **C2 Servers**: These are remote servers controlled by the attacker, used to send commands and receive data from the infected machines. They act as the central hub for managing the compromised network.

  - **Domain Generation Algorithms (DGAs)**: DGAs generate numerous domain names for C2 communication. This technique makes blocking all potential C2 channels difficult for defenders, as the malware can switch between domains dynamically.

- **Scripting and Automation**: Automation tools and scripting languages are extensively used to streamline the installation and execution of malicious payloads:

  - **PowerShell**: This powerful scripting language is often used to download and execute malicious payloads. Its versatility and deep integration with Windows systems make it a preferred tool for attackers.

  - **VBScript and Batch Files**: These scripting languages are used to automate various aspects of the malware installation process. They can execute commands, manipulate files, and interact with the operating system seamlessly.

- **Exploitation Frameworks**: Exploitation frameworks provide the necessary tools to exploit vulnerabilities and deliver payloads effectively:

  - **Exploit Kits**: Kits like RIG and Neutrino are designed to exploit known vulnerabilities in software and deliver malicious payloads. They are often used in drive-by download attacks to compromise systems silently.

  - **Custom Exploits**: Tailored specifically for the target environment, custom exploits are developed to take advantage of unique vulnerabilities. These bespoke tools are often more effective and harder to detect.

- **Encryption and Obfuscation**: To avoid detection and analysis, attackers use encryption and obfuscation techniques extensively:

  - **Encrypted Payloads**: By encrypting the payload, attackers can hide its content from security software. Only when decrypted by the malware does the payload become executable.

  - **Code Obfuscation**: Techniques such as code obfuscation make the malicious code difficult to read and analyze. This includes renaming variables, altering control flow, and inserting meaningless code.

- **Anti-Detection Techniques**: To evade detection by security measures, attackers employ various anti-detection techniques:

  - **Antivirus Evasion**: Techniques and tools are used to avoid detection by antivirus software. This includes using packers, crypters, and altering malware signatures.

  - **Sandbox Evasion**: Attackers use methods to detect and evade sandbox environments used for malware analysis. This ensures their payloads execute only in real environments, reducing the risk of detection.

By leveraging these tools and technologies, attackers ensure their malicious payloads are successfully installed, remain undetected, and maintain persistent access. This critical stage of the Cyber Kill Chain sets the foundation for further exploitation and control over compromised systems, highlighting the need for robust security measures to counteract these sophisticated threats.

Droppers and payload installation

In the *Installation* phase of the Cyber Kill Chain, attackers utilize various techniques to install payloads onto target systems. Central to this process are "droppers" – mechanisms employed to deliver and execute malicious payloads. This section delves into the role of droppers and the methods used for payload installation.

Droppers play a crucial role in the exploitation phase of cyberattacks, enabling the delivery and execution of malicious payloads. Understanding how droppers operate and the methods used for payload installation is essential for organizations to enhance their defense mechanisms and safeguard against cyber threats.

Droppers, integral to the Cyber Kill Chain, serve as vital components in cyberattacks. They act as intermediaries, bridging the gap between the attacker-controlled infrastructure and the intended target system. Their primary purpose is to facilitate the delivery and execution of malicious payloads. To achieve this, droppers are equipped with a range of evasion techniques, including encryption, obfuscation, and anti-analysis measures.

Droppers play a pivotal role in the initial stages of a cyberattack. They are the first point of entry into the target system, serving as the gateway for the attacker to establish a foothold. This foothold is crucial for the subsequent phases of exploitation and control.

Evasion techniques employed by droppers are vital for their success, as they are designed to bypass and outsmart security measures, such as antivirus software and intrusion detection systems. These techniques make it challenging for defenders to detect and thwart dropper activity effectively.

Types of droppers

Droppers come in a variety of forms, each tailored to specific objectives and the need for stealth. The following are some common types of droppers:

1. **Executable droppers**: Executable droppers are standalone programs that are capable of directly executing malicious code on the target system. They are often used when rapid and direct action is required, making them suitable for deploying payloads that must execute immediately upon infiltration.

2. **Script-based droppers**: These droppers are typically written in scripting languages like JavaScript or PowerShell. They do not directly contain the malicious payload but are responsible for downloading and executing it from remote servers or other sources. Script-based droppers are favoured for their flexibility and adaptability.

3. **Document-based droppers**: Malicious payloads can be hidden within seemingly innocuous documents, such as PDFs or Microsoft Office files. These documents exploit vulnerabilities in the associated software to deliver payloads covertly. Document-based droppers are effective in evading detection because they leverage users' trust in commonly used file formats.

The choice of dropper type depends on the attacker's specific objectives and the vulnerabilities they seek to exploit. For example, an attacker aiming for a zero-day exploit might opt for a document-based dropper to exploit the trust users place in familiar file formats.

Script-based droppers have gained prominence in recent years due to their adaptability and effectiveness in evading detection. They can be easily modified and customized to fit the attacker's needs, making them a versatile choice for delivering payloads.

Techniques for payload installation

This section will describe how attackers use different approaches to deliver and execute malicious payloads on target systems, emphasizing the importance of understanding these techniques for effective defense:

1. **Exploiting software vulnerabilities**: Attackers often use the exploitation of software vulnerabilities as a primary method to gain a foothold on target systems. This approach involves identifying and capitalizing on weaknesses in various aspects of a target system, including software applications, operating systems, and even firmware. Notably, zero-day vulnerabilities hold a special allure for attackers because they are undisclosed to the software vendor and, thus, lack available patches. The infamous Stuxnet malware campaign serves as a prime example of the successful exploitation of zero-day vulnerabilities to compromise industrial control systems.

   The exploitation of software vulnerabilities is a foundational tactic employed in numerous advanced and persistent cyberattacks. These vulnerabilities provide a gateway for attackers to enter a target system.

   Zero-day vulnerabilities are particularly dangerous because they represent uncharted territory for defenders. Without available patches, they are attractive targets for attackers, as they provide a window of opportunity for infiltration and exploitation.

2. **Drive-by downloads**: Drive-by downloads are a stealthy method employed by attackers to deliver malicious payloads to users' systems. These attacks occur when unsuspecting users visit websites that host hidden malicious content. Attackers exploit vulnerabilities in users' web browsers to automatically download and execute malicious payloads. This method is highly dynamic and can lead to widespread infections.

   Drive-by downloads are often initiated without any user interaction or consent, making them a particularly stealthy vector for malware distribution. Users may not even be aware that their systems have been compromised until after the fact.

   Regular software updates and diligent patch management are critical for mitigating the risks associated with drive-by downloads. Timely updates help to close known vulnerabilities that attackers might otherwise exploit.

3. **Social engineering**: Social engineering techniques rely on manipulating human psychology to deceive individuals into executing droppers, making it a persistent and adaptable threat in cybersecurity. These tactics encompass a wide range of strategies, including sending deceptive email attachments, providing enticing links, and crafting persuasive messages that induce users to open or execute malicious files. Human error remains a significant vulnerability in cybersecurity, and attackers exploit it to their advantage.

Social engineering is a versatile method that exploits human psychology, making it a persistent and pervasive threat in the cybersecurity landscape. It evolves alongside human behavior and trends.

Security awareness training plays a crucial role in combating social engineering attempts by educating users on the risks and tactics employed by attackers. Informed users are more likely to recognize and resist social engineering ploys, reducing the likelihood of successful attacks.

Defensive strategies for malware

Detecting malware installation is a critical component of incident response, as early detection can significantly mitigate the impact of a cyber-attack. Incident response teams employ various techniques to identify malicious software on systems. Here are some of the most common methods:

- **Behavioral analysis**: Behavioral analysis involves monitoring and analyzing the behavior of program and system processes to identify anomalies indicative of malware. This technique is based on the principle that malicious software often exhibits unusual behavior patterns, such as:

  - **Unexpected network activity**: Malware may attempt to communicate with command and control (C2) servers, leading to unusual outbound traffic patterns.

  - **File and registry changes**: Malware installation often involves creating or modifying files and registry entries to achieve persistence.

  - **Process anomalies**: Unusual processes or services running on the system, especially those with high privileges, can be a sign of malware.

- **Signature-based detection**: Signature-based detection relies on identifying known patterns of malicious code. This technique involves scanning files and memory for unique signatures associated with known malware. Although effective for detecting previously identified threats, it may be less effective against new, unknown malware variants.

  - Antivirus and antimalware tools maintain databases of known malware signatures and scan systems for matches.

  - YARA rules: Customizable rules that define patterns of interest in files and processes, helping to identify specific malware families.

- **Heuristic analysis**: Heuristic analysis goes beyond signature-based detection by examining the characteristics and behaviors of files and programs to identify potentially malicious activities. This approach can detect new or modified malware that has yet to be cataloged in signature databases.

  - **Static analysis**: Examining the code and structure of files without executing them to identify suspicious characteristics.

  - **Dynamic analysis**: Running programs in a controlled environment (sandbox) to observe their behavior and identify malicious actions.

- **Network traffic analysis**: Network traffic analysis involves monitoring and analyzing network communications for signs of malicious activity. This technique can detect malware attempting to communicate with external servers or other compromised systems within the network.

  - **Intrusion Detection Systems (IDS)**: Tools like Snort and Suricata monitor network traffic for known attack patterns and anomalies.

  - **Network flow analysis**: Examining flow records (for e.g., NetFlow, IPFIX) to identify unusual traffic patterns that may indicate malware communication.

- **Endpoint detection solutions**: Antivirus Software: Antivirus software is a fundamental component of endpoint security. It scans files and programs for known malware signatures, preventing their execution. Modern antivirus solutions incorporate heuristic analysis to identify suspicious behavior, making them highly effective in mitigating emerging threats.

  - **Host-Based Intrusion Detection Systems (HIDS)**: HIDS monitor the activities on individual devices. They generate alerts when they detect potentially malicious behaviour, such as unauthorized access or unusual system activity. This real-time monitoring is instrumental in identifying and mitigating threats swiftly.

  - **Endpint Detection & Response (EDR)**: EDR solutions provide continuous monitoring and analysis of endpoint activities to detect and respond to advanced threats. These tools combine multiple detection techniques to provide comprehensive visibility into endpoint behavior.

    - **Real-time monitoring**: Continuous observation of file, process, and network activities on endpoints.

    - **Threat hunting**: Proactively searching for signs of malicious activity using advanced query languages and behavioral indicators.

- **File Integrity Monitoring (FIM)**: FIM tools track changes to critical system files and configurations, alerting incident response teams to unauthorized modifications that may indicate malware installation.

  - **Baseline comparisons**: Comparing current system states to known-good baselines to identify unauthorized changes.

  - **Real-time alerts**: Immediate notifications of changes to critical files and directories.

- **Log analysis**: Analyzing system and application logs can reveal indicators of compromise (IOCs) associated with malware installation. Logs provide valuable insights into system activities and events that can help identify malicious actions.

  - **Security Information and Event Management (SIEM)**: Centralized platforms that aggregate and analyze log data from various sources to detect and respond to security incidents.

  - **Correlation rules**: Defining rules that correlate log events to identify patterns indicative of malware activity.

- **Memory forensics**: Memory forensics involves analyzing the contents of a system's memory to identify signs of malware that may not be visible on disk. This technique can uncover advanced threats, such as file-less malware, that reside solely in memory.

  - **Volatility Framework**: An open-source tool for performing memory forensics and analyzing volatile memory artifacts.

  - **Memory dumps**: Capturing and analyzing memory dumps to identify malicious processes and code injections.

- **User and Entity Behavior Analytics (UEBA)**: UEBA solutions analyze the behavior of users and entities within an organization to detect anomalies that may indicate malicious activity. These tools use machine learning to establish baselines and identify deviations.

  - **Behavioral baselines**: Establishing standard behavior patterns for users and entities.

  - **Anomaly detection**: Identifying deviations from established baselines that may indicate compromise.

By combining these techniques, incident response teams can effectively detect malware installations, enabling timely intervention and minimizing the impact of cyber attacks.

Notable incidents involving malware installation

Let's take a look at some case studies of malware installation to understand the real-world implications and the effectiveness of various defensive strategies.

Case Study 1: Ryuk Ransomware (2020)

Ryuk ransomware, a formidable malware strain known for its precise targeting and high ransom demands, made significant headlines in 2020 with a series of attacks directed at healthcare organizations. The incident underscored the evolving tactics and threats in the cybersecurity landscape, particularly during a global health crisis.

In 2020, healthcare providers worldwide faced immense challenges due to the COVID-19 pandemic. Hospitals and healthcare facilities were at the forefront of addressing the health crisis, making them vital and, unfortunately, lucrative targets for cybercriminals.

The consequences of Ryuk ransomware attacks in the healthcare sector were severe and far-reaching:

Ryuk attacks disrupted the normal operations of healthcare facilities, often causing delays in patient care and creating significant operational hurdles. Systems used for patient management, laboratory tests

Healthcare organizations affected by Ryuk ransomware had to divert resources to deal with the attack, which had an impact on their financial stability. In many cases, substantial ransoms were demanded in cryptocurrencies, further straining the financial resources of these organizations.

Ryuk's encryption methods rendered sensitive patient data and medical records inaccessible. This not only hindered healthcare providers' ability to deliver care but also raised concerns about patient privacy and data breaches.

The disruption of healthcare services due to ransomware attacks raised serious concerns about patient safety. Delayed access to critical medical records and treatment information could have life-threatening consequences.

Incidents like the Ryuk attacks erode public trust in the healthcare sector's ability to safeguard sensitive patient information. The reputational damage can be long-lasting.

The U.S. Department of Health & Human Services (HHS) issued an alert specifically addressing the Ryuk ransomware activity targeting the healthcare and public health sector. This alert provided valuable insights into the incident and recommendations for mitigating the impact of ransomware attacks on the healthcare industry

The Ryuk ransomware attacks in the healthcare sector brought to the forefront the urgent need for robust cybersecurity measures and preparedness. Healthcare providers and organizations took several measures to mitigate the impact of such attacks:

- **Data backups**: Organizations reinforced their data backup systems to ensure the availability of critical patient information in the event of a ransomware attack. Regular backups and off-site storage became essential.

- **Incident response plans**: Healthcare organizations developed and updated their incident response plans to address the specific challenges posed by ransomware attacks. These plans outlined procedures for containing incidents, communication, and recovery efforts.

- **Employee training**: Employee training programs were expanded to raise awareness about the risks of phishing emails and the importance of recognizing potential threats. Staff members were encouraged to report any suspicious activities promptly.

- **Advanced security measures**: Healthcare providers adopted advanced security solutions such as intrusion detection systems (IDS) and intrusion prevention systems (IPS) to monitor network traffic for signs of malicious activity.

- **Collaboration with law enforcement**: Healthcare organizations collaborated with law enforcement agencies to facilitate investigations into ransomware attacks and to potentially identify and apprehend threat actors.

- **Enhanced patch management**: Keeping software and systems up to date with security patches became a priority, as unpatched vulnerabilities often serve as entry points for ransomware attacks.

The Ryuk ransomware attacks targeting healthcare organizations in 2020 brought into sharp focus the vulnerability of critical sectors during a global health crisis. These attacks had significant consequences, from operational disruptions to financial losses, highlighting the urgent need for robust cybersecurity measures in the healthcare sector. It serves as a sobering reminder of the impact ransomware can have on public safety and the critical importance of cybersecurity in safeguarding essential services.

Case Study 2: Conti Ransomware (2021)

The Conti ransomware group, a relatively recent but increasingly prominent threat actor, gained notoriety in 2021 for its audacious attacks on critical infrastructure, with a particular focus on healthcare and emergency services. This criminal group, like many others, employed ransomware as a tool for extortion. The year 2021 witnessed a surge in their activities as they demanded substantial ransoms from their victims, bringing the threat to the forefront of the cybersecurity landscape.

Conti is part of a new ransomware breed, often called **Ransomware-as-a-Service (RaaS)**. It operates as an affiliate program, providing the ransomware code to different groups who carry out the attacks . This decentralized approach allows the developers to collect a portion of the ransom payments while insulating themselves from direct law enforcement action.

Conti is known for employing various tactics to infiltrate organizations. The initial infection often involves phishing emails or exploiting vulnerabilities in remote desktop services, VPNs, or software applications. Once inside a network, Conti deploys its ransomware, encrypting files and demanding ransoms for decryption keys.

One of the most alarming aspects of Conti's operations is its deliberate targeting of critical infrastructure providers. In particular, they have shown a penchant for attacking healthcare organizations and emergency services. These sectors are of utmost importance, and any disruption can have life-threatening consequences. Conti's choice of targets underscores the growing audacity of ransomware groups in pursuing high-profile victims.

The consequences of Conti's actions in 2021 were dire. By targeting critical infrastructure, they put lives at risk and disrupted essential services. Here are some specific instances of the impact:

- **Healthcare disruption**: Conti's attacks on healthcare providers resulted in the interruption of medical services and patient care. Hospitals faced critical situations where they had to divert patients or cancel surgeries due to the loss of access to critical patient data and systems.

- **Ransom demands**: Conti demanded substantial ransoms from their victims, often running into millions of dollars. These demands put significant financial pressure on the targeted organizations.

- **Operational downtime**: As organizations grappled with the aftermath of the ransomware attacks, there was considerable operational downtime. This impacted the efficiency and functionality of the targeted institutions.

- **Recovery costs**: Recovering from a Conti ransomware attack is not a straightforward process. It requires financial investments in forensic analysis, cybersecurity improvements, and, in some cases, the payment of the ransom itself. All of this contributes to a considerable financial burden.

- **Legal and regulatory implications**: Ransomware attacks often trigger legal and regulatory consequences, especially when personal or sensitive data is compromised. Organizations must navigate data breach notifications, legal investigations, and potential fines.

- **Heightened security awareness**: These attacks have heightened security awareness among critical infrastructure providers. While a defensive response can be costly, it is an imperative investment to prevent future attacks.

The Conti ransomware group's 2021 activities underscore the urgent need for proactive cybersecurity measures to safeguard critical infrastructure. These attacks revealed the vulnerability of essential services and the potentially life-threatening consequences of ransomware incidents. Cybersecurity practitioners, policymakers, and organizations must work collaboratively to protect critical infrastructure and ensure the safety and well-being of the public. The case of Conti serves as a stark reminder of the critical importance of robust cybersecurity measures in the face of evolving threats.

Security measures to prevent unauthorized installations

In the realm of cybersecurity, preventing unauthorized software installations is paramount to safeguarding systems and data. This section explores key security measures, such as application whitelisting and endpoint security, which are instrumental in preventing unauthorized software installations. Let's look at those measures.

- **Understanding unauthorized installations** Before diving into the specific measures, it is important to understand the context and significance of preventing unauthorized software installations. Unauthorized software can introduce vulnerabilities and increase the risk of cyber attacks. Implementing security measures like application whitelisting and endpoint security solutions is essential for maintaining a secure environment.

  - **The risks of unauthorized software**: Unauthorized software installations can introduce significant risks to an organization. Such software may contain vulnerabilities or malicious code, potentially leading to security breaches or system compromise.

  - **Attack vectors**: Attackers often exploit unauthorized software installations as attack vectors. For example, they may plant malware within seemingly innocuous applications, exploiting the trust that users place in authorized software.

- **Security measures:**Security measures like application whitelisting and endpoint security solutions play a crucial role in preventing unauthorized software installations. These measures help control which applications can run on a system, reducing the risk of malicious software being executed.

  - **Application whitelisting**: Application whitelisting is a proactive security measure that allows organizations to specify which software applications are permitted to run on their systems. This approach blocks unauthorized software from executing

    - **Benefits of application whitelisting**: Application whitelisting reduces the attack surface by allowing only trusted software to execute. It enhances security by minimizing the risk of unapproved software introducing vulnerabilities.

- **Challenges in Implementation**: While effective, application whitelisting can present challenges, particularly in environments with diverse software needs. Maintaining whitelists and addressing legitimate software updates can be complex.

- **Endpoint security solutions** Endpoint security solutions include features like behavioral analysis, which can detect and prevent suspicious software activities. These solutions provide a comprehensive approach to securing endpoints against unauthorized installations.

  - **Behavioural analysis**: Endpoint security solutions employ behavioural analysis to detect suspicious activities, including attempts to install unauthorized software. By analysing software behaviour, these solutions can proactively identify and block threats.

  - **Integrated threat intelligence**: Some endpoint security solutions leverage threat intelligence feeds to stay updated on emerging threats. This proactive approach ensures that the system can identify and mitigate unauthorized installations promptly.

Preventing unauthorized software installations is a critical aspect of cybersecurity. Security measures such as application whitelisting and endpoint security bolster an organization's defences by limiting the execution of unapproved software. Understanding and implementing these measures are essential steps in safeguarding systems and data from potential threats.

Incident response and recovery after malware installation

When faced with the unfortunate event of a malware installation, your proactive approach with a robust incident response and recovery plan is not just important, it's crucial. This approach is key to minimizing damage and restoring normal operations. The first step in this process, rapid incident detection, is of utmost importance. It involves methods and technologies that enable you to swiftly identify the presence of malware. Once detected, immediate containment strategies must be implemented to prevent the malware from spreading further within the network. This could involve isolating affected systems, turning off certain network functions, or even shutting down parts of the network to stop malware propagation.

Following containment, the focus shifts to eradication and recovery. Eradication involves thoroughly removing the malware from all affected systems, which may require reimaging infected machines, applying patches, or using specialized malware removal tools. Recovery steps include restoring affected services and data from backups and ensuring that systems are clean and secure before returning them online. This stage is not just critical, it's a beacon of hope for resuming normal business operations with minimal downtime and data loss.

Post-incident analysis is an essential component of the response and recovery process. By reviewing the incident, organizations can gain insights into the attack vector and the effectiveness of their response. This analysis helps in identifying any weaknesses in the current defenses and developing strategies to prevent future incidents. But remember, having a dedicated incident response team is not just a recommendation, it's a necessity. This team, equipped with the necessary skills and tools, is your support system during a crisis, ensuring that all members understand their roles and responsibilities and making you feel secure in the face of a cyber threat.

In summary, incident response and recovery after malware installation involve a series of coordinated steps: rapid detection, effective containment, thorough eradication, and strategic recovery. Post-incident analysis further strengthens defenses, preparing the organization for future threats. By emphasizing preparation and having a robust incident response plan, organizations can significantly reduce the impact of malware installations and enhance their overall cybersecurity resilience. A robust incident response plan is not just a document, it's a lifeline in the event of a cybersecurity incident.

The role of artificial intelligence and machine learning in preventing malware installation

In the rapidly evolving cybersecurity landscape, artificial intelligence (AI) and machine learning (ML) have emerged as powerful tools, empowering cybersecurity professionals in the fight against malware installation. These technologies offer advanced capabilities that significantly enhance the detection and prevention of malicious activities, enabling professionals to stay ahead of increasingly sophisticated cyber threats. AI and ML algorithms excel at analyzing vast amounts of data at high speed, identifying patterns and anomalies that might indicate an attempted malware installation. Unlike traditional signature-based detection methods, which rely on known patterns of malicious code, AI and ML can recognize previously unseen threats by learning from data and improving their detection models over time.

One of the primary advantages of using AI and ML in cybersecurity is their proactive ability to provide real-time threat detection and response. Machine learning algorithms can continuously monitor network traffic, system behaviors, and user activities, flagging suspicious actions deviating from established norms. This proactive approach, which is a significant departure from reactive methods, allows for swiftly identifying and mitigating threats before they can cause significant harm. For instance, AI-driven tools can detect unusual patterns in file access or system modifications that may indicate an ongoing installation attempt, triggering automated responses to isolate and neutralize the threat.

Numerous case studies have demonstrated the effectiveness of AI and ML in preventing malware installation. For example, companies deploying AI-based security solutions have reported significant reductions in successful attacks, thanks to the ability of these systems to adapt to new threats quickly. This adaptability not only instills confidence in the resilience of these technologies but also reassures decision-makers about their investment in AI and ML. They can also help in reducing false positives, which are common in traditional security systems, by providing more accurate threat assessments based on continuous learning and contextual analysis.

Looking forward, the role of AI and ML in cybersecurity is expected to grow even more critical. As cyber threats become more complex and voluminous, the scalability and efficiency of AI and ML systems will be indispensable. This future growth instills optimism about the future of cybersecurity. Future advancements may include more sophisticated AI models capable of predicting and preempting attacks based on a comprehensive understanding of an organization's unique threat landscape. However, it is also important to recognize the limitations and challenges associated with these technologies, such as the need for large datasets to train models and the potential for adversarial attacks aimed at deceiving AI systems.

Integrating artificial intelligence and machine learning into cybersecurity strategies represents a significant leap forward in preventing malware installation. By leveraging these technologies, organizations can achieve a higher level of protection, responding to threats with unprecedented speed and accuracy. As AI and ML continue to evolve, they will undoubtedly play a crucial role in shaping the future of cybersecurity, making it imperative for organizations to embrace these advancements to safeguard their digital assets effectively.

Summary

This chapter explores the *Installation* stage within the Cyber Kill Chain, a critical phase where attackers establish and maintain a foothold in compromised systems. It begins with an introduction to the importance of this stage in the lifecycle of a cyber attack. The *Method of Installation* section describes various techniques attackers use to install malware, ranging from exploiting vulnerabilities and leveraging social engineering tactics to using sophisticated tools to evade detection. The chapter delves into the common tools and technologies used in the Installation phase of the Cyber Kill Chain, covering **Remote Access Trojans (RATs)** like DarkComet and NanoCore, backdoors via frameworks such as Metasploit and Cobalt Strike, and the use of packers and crypters to obfuscate malicious payloads. Understanding these methods and tools is crucial for developing effective defensive measures.

The chapter examines how initial payloads are delivered and executed through droppers, often bypassing initial defenses. Defensive strategies are explored, offering practical measures to prevent and mitigate installation attempts. Real-world examples of significant cyber-attacks are provided in the *Notable Incidents Involving Malware Installation* section, highlighting the impact and complexity of malware installations and emphasizing lessons learned and strategies for improvement. Throughout the chapter, the focus remained on equipping readers with the knowledge needed to combat malware installation effectively. The next chapter will focus on *Command and Control*, exploring how attackers maintain communication with compromised systems and strategies to detect and disrupt these activities.

7

Command and Control

The **Command and Control (C2)** phase is not just pivotal, it's the critical juncture where attackers solidify their grip on compromised systems. This chapter delves deep into the shadowy world of C2, illuminating the techniques, strategies, and infrastructure that threat actors employ to orchestrate their malicious campaigns.

As we peel back the layers of C2 operations, we'll explore the following:

- The fundamental concepts that underpin C2 frameworks
- Sophisticated evasion techniques that keep attackers hidden in plain sight
- The role of proxy servers in obfuscating malicious traffic
- Cutting-edge defensive measures to thwart C2 activities
- The anatomy of C2 servers and their communication protocols
- Real-world case studies that bring theory into sharp focus
- Advanced detection and disruption strategies to sever the attacker's lifeline

By the end of this chapter, you'll possess a comprehensive understanding of C2 operations, a knowledge that will not just inform but also transform your approach to cybersecurity. Whether you're a seasoned security professional or an aspiring cyber defender, the insights gained here will prove invaluable in the ongoing battle against sophisticated cyber threats. This knowledge will inspire you to continuously improve and adapt in the face of evolving threats.

Prepare to embark on a thrilling journey through the nerve center of modern cyberattacks. The knowledge you gain will be the difference between a successful breach and a thwarted attempt. Let's begin our exploration of the complex and fascinating world of command and control.

Understanding the C2 phase

The C2 phase represents a pivotal stage in the cyber kill chain where cyber adversaries establish clandestine communication channels with compromised systems. This phase enables them to exert control, issue commands, exfiltrate sensitive data, and sustain their presence within the compromised environment. To fully comprehend the intricacies of the C2 phase, it is essential to delve into the techniques and strategies employed by malicious actors. The following diagram shows a step-by-step breakdown of the C2 phase:

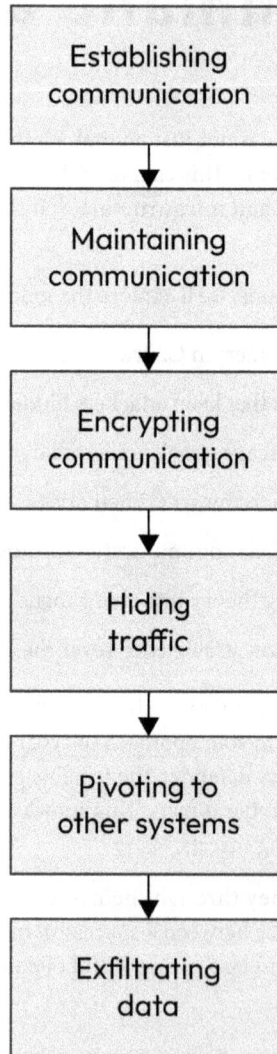

```
┌─────────────────────────┐
│     Establishing        │
│     communication       │
└─────────────────────────┘
            │
            ▼
┌─────────────────────────┐
│     Maintaining         │
│     communication       │
└─────────────────────────┘
            │
            ▼
┌─────────────────────────┐
│     Encrypting          │
│     communication       │
└─────────────────────────┘
            │
            ▼
┌─────────────────────────┐
│       Hiding            │
│       traffic           │
└─────────────────────────┘
            │
            ▼
┌─────────────────────────┐
│     Pivoting to         │
│    other systems        │
└─────────────────────────┘
            │
            ▼
┌─────────────────────────┐
│     Exfiltrating        │
│       data              │
└─────────────────────────┘
```

Figure 7.1 – Techniques and strategies for establishing communication channels

Let's go through each strategy one by one:

- **Establishing communication**: Attackers, driven by the imperative to remain concealed and avoid detection, employ many covert techniques to establish communication channels with compromised systems. These channels are meticulously designed to blend seamlessly with legitimate network traffic, making it exceptionally challenging for network defenders to identify and thwart their activities. The practice is too common in the cyber threat landscape, underscoring the prevalence of covert communication channels.

- **Maintaining communication**: The covert channels are designed to mimic regular network traffic patterns, often leveraging common communication protocols such as HTTP, DNS, **Internet Relay Chat** (**IRC**), and even peer-to-peer networks. The channel choice depends on the attacker's specific objectives, target environment, and the need for discretion. Each of these channels has unique characteristics that attackers exploit to their advantage. For instance, HTTP, being a widely used protocol, provides a convenient cover for malicious communication, allowing malicious traffic to blend in with legitimate web traffic. On the other hand, peer-to-peer networks are decentralized, making them harder to trace. DNS, an essential service for internet communication, is often leveraged due to its ubiquity and potential for bypassing traditional security measures.

- **Encrypting communication**: Attackers frequently encrypt their communications to secure their communication channels further and ensure the integrity and confidentiality of the data being transmitted. This encryption makes it significantly more difficult for network defenders to detect and analyze malicious traffic, as it appears gibberish to unauthorized observers.

- **Hiding traffic**: In addition to encrypting communications, attackers take measures to hide their traffic from detection tools and techniques. This can involve using obfuscation methods, such as encoding data to make it less recognizable or embedding malicious traffic within seemingly innocuous data streams.

- **Pivoting to other systems**: Once communication is established and maintained, attackers often seek to pivot to other systems within the compromised network. This lateral movement allows them to expand their control, gain access to additional resources, and further entrench themselves within the target environment.

- **Command and data exfiltration**: In addition to establishing communication and issuing commands, attackers frequently employ the C2 channel to exfiltrate sensitive data from compromised systems. Data exfiltration is a critical aspect of their operation, as it enables them to steal valuable information without alerting the target organization. To maintain their covert status, attackers often exfiltrate data in small, inconspicuous chunks, thereby avoiding suspicion. This method, known as **low and slow** data exfiltration, is particularly effective at eluding detection, as it does not trigger significant spikes in network traffic that could raise alarm bells.

In the following sections, we will explore the various methods attackers use to evade detection during the C2 phase, including encryption, obfuscation, and traffic manipulation techniques.

Evading detection in the C2 phase

In the cat-and-mouse cybersecurity game, the C2 phase is where attackers test their stealth skills. This critical stage involves maintaining hidden communication channels with compromised systems while evading detection. Cybercriminals employ sophisticated techniques to disguise their activities, from encryption and obfuscation to traffic manipulation. By understanding these evasion tactics, cybersecurity professionals play a crucial role in defending against persistent threats and minimizing the impact of potential breaches. Let's explore the clever methods attackers use to stay hidden and how these professionals can spot these digital ghosts in the machine:

- **Encryption and obfuscation**: During the C2 phase, attackers employ sophisticated encryption and obfuscation techniques as part of their arsenal to conceal the malicious nature of their communication. These methods are strategically designed to make it extremely challenging for security tools and analysts to discern their activities.

- **Custom encryption algorithms**: Attackers often create custom encryption algorithms to encrypt their communication. These algorithms are not publicly available, making it difficult for defenders to decrypt the traffic without prior knowledge of the algorithm's specifics.

- **Encoding schemes**: Encoding schemes, such as Base64 encoding or binary encoding, are frequently used to obscure the content of C2 traffic. By converting their commands and data into a different format, attackers aim to bypass signature-based detection methods that rely on known patterns.

- **Traffic pacing**: Attackers may employ techniques to vary the timing and volume of C2 traffic, mimicking legitimate network behavior. This irregular pacing makes it harder for intrusion detection systems to identify patterns associated with malicious communication.

- **Payload encryption**: Beyond encrypting the entire communication channel, attackers can selectively encrypt specific payloads within the traffic. This enables them to send critical commands or exfiltrate sensitive data without raising suspicion.

- **Dynamic keys**: Attackers may change encryption keys dynamically during the communication session, adding another layer of complexity for defenders. This dynamic key exchange makes it difficult to intercept and decrypt the data in real time.

Use of proxy servers

Attackers frequently leverage proxy servers to further obfuscate the origin and destination of their C2 traffic. This method adds an additional layer of complexity for defenders, making it challenging to trace the source of the malicious communication. The following points highlight various techniques involving proxy servers:

- **Anonymizing source IPs**: By routing C2 traffic through proxy servers, attackers can hide their actual source IP addresses. This practice makes it appear as though the traffic is originating from the proxy server, effectively masking their identity.

- **Geographical diversification**: Attackers often use proxy servers located in various geographic regions, making it difficult to pinpoint the geographical source of the C2 traffic. This diversification is particularly useful for evading geographic-based filtering and analysis.

- **Proxy chaining**: Some advanced attackers employ a technique known as proxy chaining, where traffic is routed through multiple proxy servers in a chain. Each proxy in the chain forwards the traffic to the next, making it exceedingly complex for defenders to trace the path of the communication.

- **Tor and anonymity networks**: In some cases, attackers route C2 traffic through anonymity networks such as Tor, which add layers of encryption and routing through a distributed network of volunteer-run servers. This makes the source of the communication nearly impossible to determine.

Defensive measures

Effective defensive measures are essential for detecting and mitigating C2 activities. The following strategies can help organizations protect against C2 operations:

- **Network traffic analysis**: Organizations can implement network traffic analysis solutions to monitor and analyze network traffic for unusual patterns indicative of C2 communication.

- **Behavioral analysis**: Employing behavioral analysis tools can help identify abnormal system behavior caused by C2 communication, triggering alerts for investigation.

- **Threat intelligence feeds**: Subscribing to threat intelligence feeds can provide organizations with real-time information about known C2 infrastructure, aiding in the detection and blocking of malicious communication.

The C2 phase is a critical component of the cyber kill chain, where attackers establish communication with compromised systems. Understanding the techniques employed by attackers in this phase and implementing appropriate defensive measures is essential for organizations to detect, mitigate, and respond effectively to cyber threats.

By understanding these defensive measures, organizations can better prepare for the challenges posed by C2 activities. In the following sections, we will delve deeper into the mechanics of C2 servers and their communication protocols.

C2 servers and communication

C2 servers play a pivotal role in the execution of cyberattacks, serving as the central communication hub for threat actors. In this section, we will explore the mechanics of C2 servers and how they facilitate cyberattacks.

C2 servers serve as the linchpin of cyberattacks, enabling threat actors to maintain control over compromised systems and execute malicious operations. Understanding the mechanics of C2 communication is vital for organizations to detect, respond to, and mitigate cyber threats effectively.

Understanding C2 servers

C2 servers are the clandestine architects of cyberattacks, wielding immense power and influence in the digital realm. At the core of their operation is their pivotal role as a nexus, linking malevolent actors with compromised systems. This link empowers attackers with the ability to remotely manipulate and communicate with their compromised assets, thereby allowing them to execute commands, surreptitiously exfiltrate valuable data, and establish an enduring foothold within targeted networks. The comprehension of this concept is indispensable in deciphering the intricacies of the cyber kill chain, as it represents the very stage where malefactors craftily initiate and retain dominion over the infrastructure of their victims.

C2 servers operate as a critical pivot point within the realm of cyber warfare. They serve as the bridgehead for attackers, enabling them to exert control and facilitate communication with compromised systems. This control encompasses a broad spectrum of activities, ranging from data extraction to the execution of malicious commands. For example, attackers can instruct compromised systems to exfiltrate sensitive corporate data or launch further attacks against additional targets. The stealth and resilience of C2 infrastructure are paramount for the success of these malicious operations.

Stealth and resilience are hallmarks of C2 server design, embodying the sophistication and dedication of modern threat actors. To achieve stealth, attackers employ **domain generation algorithms** (**DGAs**), a technique that generates dynamic and ever-changing domain names. The dynamic nature of these domains poses a formidable challenge to defenders, rendering it arduous to pre-emptively block or discern malicious communication. DGAs allow attackers to continually adapt, making it difficult for defenders to predict their movements or anticipate their next steps. This adaptability significantly contributes to the effectiveness of C2 servers.

Encryption and obfuscation further enhance the obscurity of C2 communications. Attackers meticulously employ various encryption protocols and encoding techniques to cloak their activities from the prying eyes of cybersecurity measures. By encrypting their communications, they ensure that even if intercepted, the content remains indecipherable. This practice serves as a critical defensive measure to evade detection, as it effectively masks the intent and substance of the exchanged data. Such methods require cybersecurity professionals to invest substantial effort and expertise in decryption and analysis to unveil the malicious payloads hidden within C2 traffic.

The anatomy of C2 communication

C2 communication is the lifeline of sophisticated cyberattacks, allowing threat actors to manage compromised systems remotely. This covert channel enables malware control, data theft, and orchestration of further attacks. C2 infrastructure is designed for resilience and stealth, using encryption, obfuscation, and traffic manipulation to avoid detection. By understanding C2 communication—from channel establishment to operational protocols—we gain crucial insights into modern cyber threats and can develop more effective defense strategies.

The following points outline key aspects of C2 communication that are crucial for understanding how attackers manage compromised systems:

- **Beaconing**: One of the defining features of C2 communication is beaconing. In the cyber landscape, beaconing refers to the process where compromised systems establish periodic connections with a C2 server. These connections serve the purpose of receiving instructions or reporting their status to the attacker. Beaconing is akin to a lifeline that enables the attacker to maintain a continuous channel of communication with the compromised network. Beaconing is of paramount importance to cyber criminals, as it ensures that their presence remains hidden and undetected within the compromised network. By sending intermittent beacons, the attacker can evade network monitoring tools that often focus on sustained or anomalous traffic patterns. This method of communication acts as a cloak that obscures malicious activity, allowing the attacker to control and manipulate the network from the shadows.

- **DGAs**: DGAs are sophisticated algorithms used by malware to generate many potential domain names. Attackers utilize DGAs to stay one step ahead of security measures and establish communication channels with compromised systems. The rationale behind this technique is to evade network defense mechanisms that typically rely on blacklisting known malicious domains.

  DGAs create a substantial pool of potential domains, making it challenging for defenders to predict which domains the attacker will register for C2 communication. Attackers register only a subset of these domains, making it nearly impossible for network defenders to pre-emptively block malicious domains.

  The constant evolution of DGAs creates a perpetual cat-and-mouse game between attackers and defenders. As defenders develop countermeasures to detect and mitigate DGAs, attackers respond with increasingly complex and adaptive algorithms.

- **HTTP(S) communication**: Many C2 communication channels leverage the use of common web protocols, specifically HTTP and HTTPS. This strategy is an example of attackers camouflaging their activity amidst legitimate network traffic. By using web protocols, malicious communication becomes difficult to distinguish from benign network activity.

  HTTP(S) communication channels have the advantage of traversing through network security infrastructure without raising immediate red flags. Network defenders often allow HTTP(S) traffic, as blocking it would disrupt legitimate web services. As a result, attackers utilize this to their advantage, embedding their C2 instructions within HTTP(S) requests.

  The use of encrypted HTTPS communication further complicates detection, as it conceals the content of the communication from network surveillance. As such, HTTP(S) communication represents a significant challenge for defenders seeking to identify and thwart C2 activity.

- **Data exfiltration**: The data exfiltration mechanism within a C2 server architecture involves two primary stages: data staging and data transmission. Initially, collected data is stored temporarily on the compromised system or an intermediate server, serving as a staging area before its final transfer to the C2 server. This approach ensures that data is aggregated and prepared for exfiltration without immediately alerting security mechanisms. Following this, data transmission occurs, utilizing techniques designed to securely and covertly transfer the staged data back to the C2 server. These techniques often involve using encrypted and obfuscated channels, which help avoid detection by network monitoring tools and ensure the successful extraction of valuable information from the compromised network.

C2 frameworks

C2 frameworks are essential tools in cybersecurity. Both attackers and defenders use them to manage compromised systems and understand potential attack strategies. This overview examines three prominent C2 frameworks: Cobalt Strike, Sliver, and Mythic.

Cobalt Strike is a widely-used commercial C2 framework, originally designed for red team operations but is now popular among cybercriminals and nation-state actors. It's known for its powerful features and user-friendly interface.

Key features of Cobalt Strike include the following:

- Beacon payload with multiple communication protocols
- Malleable C2 profiles for customizing traffic patterns
- Extensive post-exploitation modules
- Team collaboration support
- Scripting engine for automation and customization

Sliver is an open source, cross-platform C2 framework developed as a free alternative to commercial tools such as Cobalt Strike. It offers robust features for red team operations and is actively maintained by the cybersecurity community.

Key features of Sliver include the following:

- Cross-platform support (Windows, Linux, and macOS)
- Flexible payloads (staged, stageless, and in-memory)
- Multi-user collaboration
- Encrypted communications
- Dynamic code generation for evasion

The following image shows the Sliver Framework on Kali Linux box:

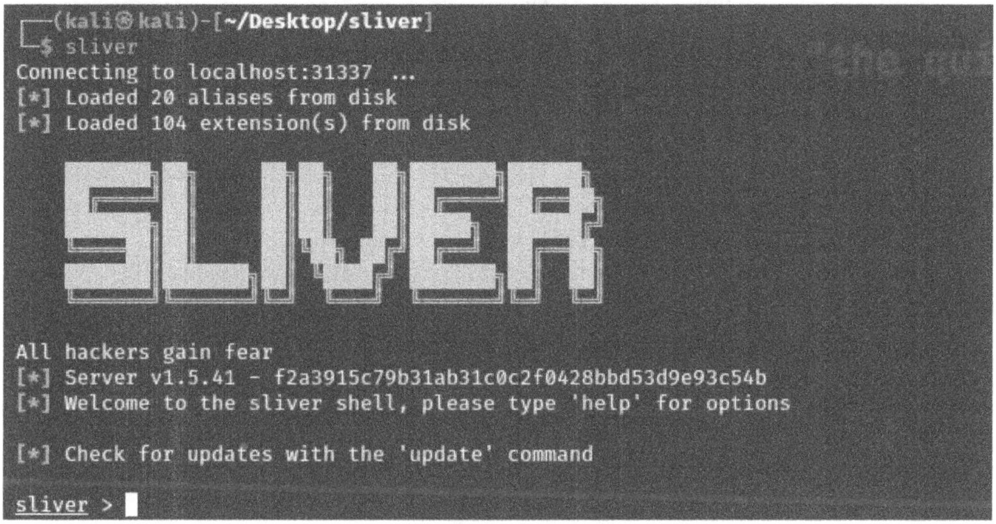

Figure 7.2 – Sliver Framework

Mythic is another open source C2 framework designed for modularity and extensibility. It uses modern web technologies and allows for extensive customization of the C2 infrastructure.

Key features of Mythic include the following:

- Modular architecture for easy customization and extension
- Web-based user interface
- Support for multiple programming languages in agents
- Secure, adaptable communication methods
- Container-based deployment for easy setup and management

These C2 frameworks represent advanced tools in the cybersecurity landscape. Their sophistication highlights the need for equally advanced detection and response strategies. Understanding these frameworks is crucial for cybersecurity professionals to anticipate and counter potential threats effectively.

Detection and mitigation

Let's learn about the command ways to detect the C2 connections:

- **Signature-based detection**: Intrusion detection systems often rely on signatures to identify known C2 traffic patterns. Signature-based detection is effective against known threats but may not catch new or custom C2 channels.

- **Behavioral analysis**: Advanced threat detection solutions use behavioral analysis to identify anomalies in network traffic patterns. This can help detect C2 communication that deviates from the norm.

- **DNS sinkholes**: Sinkholing involves redirecting C2 traffic to controlled servers, disrupting attacker control. DNS sinkholes are a common technique used to block malicious C2 domains.

- **TLS mitigation techniques**: Mitigation techniques for TLS encryption include implementing session and **deep packet inspections** (**DPI**) to detect and analyze malicious content within encrypted traffic. By deploying SSL/TLS decryption capabilities, security devices can inspect encrypted sessions for threats without compromising data integrity. Additionally, using advanced threat intelligence and machine learning models can enhance the identification of anomalies in encrypted traffic, improving the overall security posture. Properly managing TLS certificates and configurations also ensures secure and trusted communication channels, reducing the risk of exploitation.

C2 servers are at the heart of cyberattacks, enabling attackers to maintain control, exfiltrate data, and issue commands. Understanding their mechanics and the intricacies of C2 communication is essential for effective detection and mitigation of cyber threats.

Case studies illustrating C2 attacks

In this section, we will examine real-world case studies illustrating C2 attacks. These case studies provide insights into the tactics and strategies used by attackers, offering valuable lessons for cybersecurity professionals.

Case study 1 – a global botnet

Conficker, first detected in 2008, stands as one of the most notorious examples of a global botnet. It infected millions of computers worldwide, creating a massive network of compromised machines. What set Conficker apart was its utilization of a decentralized C2 infrastructure, making it incredibly challenging to disrupt. The botnet was involved in various malicious activities, including spreading malware and launching **distributed denial of service** (**DDoS**) attacks.

To appreciate the magnitude of Conficker's impact, we must acknowledge the sheer scale of its infection. Conficker managed to infiltrate millions of computers across the globe. Its ability to propagate was unprecedented, as it exploited vulnerabilities in Windows operating systems to infect new hosts. The vast botnet it established granted its operators immense power, as they could potentially harness the processing and network capabilities of millions of compromised computers.

One of Conficker's defining features was its decentralized C2 infrastructure. This mechanism allowed infected computers to communicate and receive commands from multiple sources, making them exceptionally resilient to takedown attempts. In a typical botnet, a central server controls all infected machines. In contrast, Conficker's decentralized design meant that even if one C2 server was taken down, the botnet could adapt by communicating through other infected hosts. This level of redundancy and adaptability was a significant challenge for security experts and law enforcement agencies attempting to disrupt it.

Conficker was not just an idle presence on infected machines; it actively engaged in malicious activities. One of its most concerning functions was the capability to download and execute additional malware. This made it a valuable tool for cybercriminals who could use Conficker to distribute payloads and expand their influence. Furthermore, Conficker was implicated in launching DDoS attacks, which could disrupt online services and organizations.

Dealing with Conficker was a substantial challenge due to its decentralized nature and infection scale. Detection efforts revolved around monitoring network traffic for signs of infection and identifying compromised systems. Once a Conficker-infected machine was identified, the removal process could be complex, often involving the application of security patches and comprehensive cleanup to ensure the removal of all malware components. Moreover, the creation of security patches for the specific vulnerabilities that Conficker exploited was a crucial part of the disruption process. These patches not only helped in cleaning infected systems but also served to prevent further infections.

Another disruption strategy involved sinkholing Conficker's C2 domains. Sinkholing refers to the process of redirecting traffic from malicious domains to controlled servers. By taking control of these domains, security experts were able to intercept communication from infected machines and, in some cases, send commands to Conficker-infected hosts. This method allowed for the disruption of some of Conficker's communication channels and helped reduce its influence.

Case study 2 – APT29 – Cozy Bear's persistent C2 activities

A type of **advanced persistent threat** (**APT**) group, APT29, also known as Cozy Bear, is attributed to Russian intelligence agencies. APT29 is renowned for its persistent and sophisticated C2 activities. This threat actor has targeted various organizations, including government entities, defense contractors, and diplomatic missions. APT29's C2 infrastructure includes custom malware and covert communication channels.

APT29 is characterized by its ability to maintain persistent access to compromised systems over extended periods. These actors establish covert communication channels that enable them to control infected hosts discreetly. The sophistication of their C2 infrastructure allows them to blend into the background and evade detection while continuing to carry out their espionage activities.

A key element of APT29's C2 operations is the use of custom-built malware. These malicious tools are tailored to the group's specific objectives and are designed to be elusive, making them challenging to detect using traditional security measures. The group continually evolves its arsenal of custom malware to remain ahead of security defenses.

Detecting APT29's C2 activities is a formidable challenge. The group employs advanced obfuscation techniques and encryption to conceal their communication with infected systems. Traditional signature-based detection methods are often ineffective, as APT29 is adept at avoiding well-known patterns and **indicators of compromise (IOCs)**. This necessitates the use of advanced threat-hunting techniques that focus on anomalous behaviors and patterns indicative of APT activity.

Responding to APT29's C2 activities requires a multifaceted approach. When an intrusion is detected, organizations often opt to isolate compromised systems to prevent further data exfiltration or lateral movement. The removal of APT29's custom malware is a meticulous process that involves analyzing the specific malware variant and developing countermeasures.

Strategies for detecting and disrupting C2 activities

The detection and disruption of C2 activities are critical aspects of cybersecurity, as they play a pivotal role in thwarting cyberattacks. In this section, we'll explore techniques and technologies for identifying and disrupting C2 activities, including network traffic analysis and incident response.

Understanding C2 activities

Understanding C2 activities is crucial for developing effective defense strategies. The following points highlight key aspects of C2 activities:

- **Defining C2 activities**: In the context of cybersecurity, C2 activities refer to the communication and coordination between compromised systems and the malicious actors behind cyberattacks. These activities enable attackers to remotely control compromised systems, execute malicious commands, and exfiltrate data. C2 activities are a core element of APTs and other sophisticated attacks. By establishing a secure channel for communication, attackers can maintain persistent access to the compromised system, ensuring that they can continue their malicious actions undetected. Understanding the nature and significance of C2 activities is vital for developing effective defense strategies against cyber threats.

- **Types of C2 protocols**: Attackers employ various communication protocols for C2 activities. These protocols often mimic legitimate network traffic, making them challenging to detect. Common C2 communication methods include HTTP, DNS, and IRC. For instance, malicious actors might use HTTP requests to transmit commands and receive data from compromised systems, disguising their activities within the vast amount of legitimate web traffic. Understanding these C2 communication protocols is crucial for network defenders, as it allows them to recognize and differentiate between normal and malicious network traffic. Detecting anomalous traffic patterns associated with these protocols is a key element in identifying potential C2 activities.

Techniques for detection

Effective detection techniques are essential for identifying and responding to C2 activities. The following points outline various techniques for detecting C2 activities:

- **Network traffic analysis**: Network traffic analysis is a fundamental technique for detecting C2 activities. It involves the continuous monitoring and analysis of network traffic to identify irregular patterns or behaviors that may indicate a security threat. When it comes to C2 activities, network traffic analysis helps in recognizing unusual data flows, high volumes of outbound traffic, or connections to suspicious IP addresses. This approach is particularly effective in identifying C2 activities that use specific protocols or communication patterns, as it can uncover deviations from established network baselines. By closely examining network traffic and applying anomaly detection methods, organizations can promptly detect and respond to C2 activities before significant damage occurs.

- **Behavioral analysis**: Behavioral analysis extends beyond merely monitoring network traffic and examines the behaviors of systems, applications, and users. It focuses on recognizing deviations from normal behavior that may indicate a security breach. In the context of C2 activities, behavioral analysis helps in identifying unauthorized access, lateral movement within a network, or data exfiltration. By creating behavioral profiles for network entities and continuously comparing their actions to these profiles, security teams can identify discrepancies that suggest C2 activities. For example, if a user account suddenly exhibits atypical behavior by accessing sensitive data or attempting privilege escalation, behavioral analysis can flag this as a potential C2 activity.

Technologies for detection

Various technologies can aid in the detection of C2 activities. The following points describe some of the key technologies used for this purpose:

- **Intrusion detection systems (IDSs)**: IDSs are specialized security tools designed to identify and alert organizations to potential security threats, including C2 activities. Network-based IDSs monitor network traffic, comparing it to known attack signatures and established behavioral baselines. Host-based IDSs, on the other hand, focus on the activities of individual systems or devices. IDS tools can be configured to detect known C2 communication patterns, such as connections to malicious IP addresses or the use of C2 protocols such as HTTP. When an IDS identifies a potential C2 activity, it generates alerts for security teams to investigate and take appropriate action.

- **Security information and event management (SIEM)**: SIEM systems play a vital role in C2 activity detection by collecting and analyzing log data from various sources, such as network devices, servers, and applications. These systems correlate events and log entries to detect potential C2 activities that may not be apparent when analyzed in isolation. By aggregating and analyzing logs, SIEM can identify patterns and anomalies indicative of C2 communication. SIEM's real-time alerting capabilities ensure that security teams can respond promptly to potential threats. Moreover, SIEM supports historical log analysis, enabling organizations to investigate past incidents and identify any ongoing C2 activities.

Incident response strategies

Effective incident response strategies are critical for mitigating the impact of C2 activities. The following points outline key strategies for incident response:

- **Early detection**: Early detection of C2 activities is essential for effective incident response. Identifying potential C2 activities at an early stage allows organizations to minimize the impact of cyberattacks and prevent attackers from achieving their objectives. Rapid detection often involves using a combination of network traffic analysis, IDS alerts, and SIEM data. Incident responders need to be well-versed in recognizing C2 activity indicators, including traffic patterns, protocol usage, and behavioral anomalies. Prompt action at this stage can help prevent further compromise and data exfiltration.

- **Isolation and mitigation**: Once C2 activities are detected, incident response teams must swiftly isolate compromised systems and mitigate the damage. Isolation involves disconnecting the affected systems from the network to prevent the attacker from continuing their activities and to limit lateral movement within the organization's infrastructure. Mitigation measures may include patching vulnerabilities, removing malware, and strengthening security controls. The goal is to minimize the impact of the cyberattack and prevent further unauthorized access.

- **Attribution and legal considerations**: In some cases, attributing a cyberattack to the responsible entity is essential for prosecution and legal action. This step involves identifying the attackers and gathering evidence that can be used in legal proceedings. It's important to involve law enforcement agencies and follow proper legal procedures when dealing with cybercrime. Handling evidence correctly and complying with legal requirements is crucial in building a case against malicious actors and bringing them to justice.

Summary

This chapter comprehensively explored the C2 phase, a critical component of the cyber kill chain. We examined how attackers establish and maintain covert communication channels with compromised systems, orchestrating their malicious activities from afar. The chapter covered the fundamentals of C2, detailing its purpose and structure in cyberattacks. We delved into various evasion techniques employed by threat actors, including encryption, obfuscation, stealth tactics, and the strategic use of proxy servers to mask C2 traffic. The discussion on defensive strategies highlighted proactive and reactive measures, emphasizing the importance of robust security policies, intrusion detection systems, and behavioral analysis. By dissecting C2 infrastructure and communication protocols, we provided a granular view of how attackers relay commands and exfiltrate data. Real-world case studies illustrated successful C2 operations, offering valuable lessons for cybersecurity professionals. The chapter also explored advanced detection and disruption techniques, including anomaly detection, signature-based methods, and machine learning approaches. We concluded with key takeaways and actionable recommendations, equipping you with practical knowledge to defend against C2 operations. As we transition to the next chapter, *Actions on Objectives*, cybersecurity professionals are better prepared to anticipate, detect, and disrupt this critical phase of cyber-attacks, strengthening their organization's overall security posture.

Further reading

Following are some bonus reading materials for you to check out:

- *CrowdStrike article explaining command and control attacks, how they work, and ways to defend against them*: `https://www.crowdstrike.com/cybersecurity-101/cyberattacks/command-and-control/`

- *SentinelOne overview of command and control servers, their role in cyberattacks, and detection strategies*: `https://www.sentinelone.com/cybersecurity-101/threat-intelligence/what-are-command-control-c2-servers/`

- *Fortinet glossary entry defining command and control attacks and outlining key concepts*: `https://www.fortinet.com/resources/cyberglossary/command-and-control-attacks`

- *Tripwire blog post detailing what command and control attacks are and how to protect against them*: `https://www.tripwire.com/state-of-security/what-are-command-and-control-attacks`

- *Optiv cybersecurity dictionary definition of command and control (C2)*: `https://www.optiv.com/cybersecurity-dictionary/c2-command-and-control`

- *Splunk blog explaining command and control attacks and how to prevent them*: `https://www.splunk.com/en_us/blog/learn/c2-command-and-control.html`

- *Sekoia glossary entry on command and control infrastructure used by threat actors*: `https://www.sekoia.io/en/glossary/command-control/`

- *GitHub repository for Sliver, an open-source cross-platform adversary emulation/red team framework*: `https://github.com/BishopFox/sliver`

- *Documentation for Mythic, a command and control framework for red team operations*: `https://docs.mythic-c2.net/`

Subscribe to _secpro – the newsletter read by 65,000+ cybersecurity professionals

Want to keep up with the latest cybersecurity threats, defenses, tools, and strategies?

Scan the QR code to subscribe to **_secpro**—the weekly newsletter trusted by 65,000+ cybersecurity professionals who stay informed and ahead of evolving risks.

https://secpro.substack.com

Get This Book's PDF Version and Exclusive Extras

UNLOCK NOW

Scan the QR code (or go to packtpub.com/unlock). Search for this book by name, confirm the edition, and then follow the steps on the page.

Note: Keep your invoice handly. Purchase made directly from Packt don't require one.

8

Actions on Objectives

The *Actions on Objectives* phase, the crescendo of a cyber adversary's campaign, is a crucial concept to understand. As the final stage of the cyber kill chain, it's the moment when attackers, having meticulously navigated preceding phases, execute their ultimate goals. These objectives span a broad spectrum, from data exfiltration and system sabotage to espionage and financial exploitation. Understanding these actions' motivations and methodologies is pivotal for crafting resilient defense strategies, and it's this understanding that will make you feel more knowledgeable and prepared in the face of cyber threats.

This chapter delves into the multifaceted landscape of attack objectives and techniques employed by cyber adversaries. We'll explore how these goals shape the nature and evolution of cyber threats, focusing on the far-reaching implications of economic and operational data exfiltration. The discussion will extend to effective mitigation and response strategies, with a strong emphasis on the critical role of a well-orchestrated incident response plan. This emphasis will make you feel reassured and confident in your ability to respond to cyber threats.

We'll trace the evolution of objectives-based attacks by illuminating case studies and extracting valuable insights from historical incidents. The chapter culminates in a comprehensive outline of defensive strategies tailored to counter these sophisticated attacks, offering practical guidance for fortifying systems against the advanced tactics of determined adversaries.

This chapter dissects the action on objectives phase to equip you with a deeper understanding of the endgame in cyber warfare. It fosters a proactive approach to cybersecurity in an increasingly complex threat landscape.

The following is the list of main topics covered in this chapter:

- Different types of attacks
- Cyberattack objectives and attacker profiles
- Data exfiltration
- Proactive defensive strategies to thwart objectives-based attacks
- The eighth phase of the cyber kill chain

Different types of attacks

The *Actions on Objectives* stage is the final step in the cyber kill chain, representing the attacker's ultimate goal. After cyber-criminals have developed their weapons, installed them onto a target's network, and taken control of the systems, they move on to achieve their primary objectives. These objectives can vary widely and are typically driven by the attackers' motivations, which can include financial gain, disruption of services, data theft, or long-term espionage. Understanding these objectives is crucial for developing effective defensive strategies:

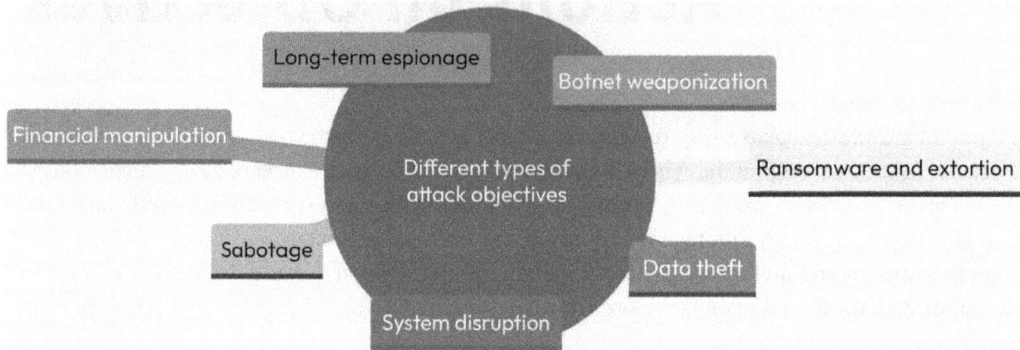

Figure 8.1 – Different types of attack objectives

Let's look into these objectives one by one:

- **Data theft**: One of the most common objectives is data theft. Attackers seek to steal sensitive information such as intellectual property, personal data, financial information, or confidential business details. This stolen data can be sold on the black market, used for identity theft, or leveraged for competitive advantage. Data theft is particularly damaging as it can lead to significant financial loss, reputational damage, and legal consequences for the affected organization.

- **Ransomware and extortion**: Ransomware attacks have become increasingly prevalent, where attackers encrypt critical data and systems, demanding a ransom for their release. The primary objective here is financial gain. Attackers use sophisticated encryption methods to lock victims out of their own systems, effectively holding their data hostage. Even if the ransom is paid, there is no guarantee that the attackers will provide the decryption key or that they have not retained copies of the data for further extortion.

- **System disruption**: Another objective is the disruption of services. Attackers may aim to incapacitate an organization's operations by launching **Distributed Denial of Service (DDoS)** attacks, which flood the target's servers with excessive traffic, rendering them unusable. Such attacks can disrupt business operations, lead to loss of revenue, and damage customer trust. The objective may be purely to cause chaos or to distract from other malicious activities happening simultaneously.

- **Long-term espionage**: In some cases, attackers may establish a foothold within a network to maintain persistent access for long-term espionage. This involves gathering intelligence over extended periods, often going undetected. The information collected can include strategic business plans, government secrets, or insider information, which can be used to gain a competitive edge, influence markets, or compromise national security.

- **Botnet weaponization**: Attackers may also aim to weaponize a botnet—a network of compromised computers controlled remotely. These botnets can be used for various malicious activities, including launching coordinated DDoS attacks, distributing malware, or conducting mass spam campaigns. The objective here is often to create widespread disruption or to monetize the botnet by offering it as a service to other cyber-criminals.

- **Financial manipulation**: In some sophisticated attacks, the objective might involve manipulating financial systems. This can include altering financial records, executing unauthorized transactions, or manipulating stock prices through false information or direct interference. Such attacks can result in significant financial losses and undermine the integrity of financial markets.

- **Sabotage**: Attackers may also have objectives aligned with causing physical or operational damage to critical infrastructure. This can involve sabotaging industrial control systems, leading to the malfunction of essential services such as power grids, water supply, or transportation networks. Such attacks can be severe, affecting public safety and national security.

By understanding the diverse objectives that attackers may have, organizations can better anticipate potential threats and develop targeted defensive strategies. Recognizing these goals allows defenders to prioritize resources and responses effectively, ensuring critical assets are protected and minimizing the impact of any successful attack.

Cyberattack objectives and attacker profiles

Comprehending the intricate relationship between attack objectives and attacker types in the complex cybersecurity landscape is not just important, it's crucial. This understanding is the key to developing robust defense strategies. Each category of cyber adversary operates with distinct goals, employing specialized techniques to achieve their aims. This mapping of objectives to attacker profiles provides a vital framework for decoding the motivations and methodologies driving cyberattacks.

Organizations can gain invaluable insights into potential threats by meticulously categorizing attack objectives and correlating them with specific attacker types. This nuanced understanding enables the development of targeted and proactive defense mechanisms. The following overview illuminates the primary attack objectives, offering detailed descriptions and identifying the attacker profiles most commonly associated with each goal.

This comprehensive analysis serves multiple purposes:

- It highlights the multifaceted nature of cyber threats in today's digital ecosystem
- It underscores the necessity for tailored defense strategies that address specific attacker motivations
- It provides a strategic foundation for risk assessment and resource allocation in cybersecurity efforts

By leveraging this mapping, security professionals are not just gaining knowledge; they're also gaining power. They can anticipate potential attack vectors, prioritize protective measures, and craft more effective incident response plans. Ultimately, this knowledge empowers organizations to stay one step ahead in the ever-evolving cyber threat landscape, safeguarding their critical assets and maintaining operational resilience.

Objective type	Description	Attacker type
Data theft	Stealing sensitive information for financial gain or competitive advantage	Cybercriminals, Nation-State Actors, Hacktivists
Ransomware and extortion	Encrypting data and demanding ransom for decryption or non-disclosure	Cybercriminals, Ransomware Gangs
System Disruption	Disrupting operations through DDoS attacks or sabotage	Hacktivists, Cyberterrorists, Cyber Vandals, Cybercriminals
Long-Term Espionage	Maintaining persistent access for ongoing intelligence gathering	Nation-State Actors, Advanced Persistent Threats (APT) Groups
Botnet Weaponization	Creating and using botnets for large-scale attacks or selling as a service	Cybercriminals, Botnet Operators
Financial Manipulation	Altering financial records or executing unauthorized transactions	Cybercriminals, Insider Threats
Sabotage	Causing physical or operational damage to critical infrastructure	Cyberterrorists, Nation-State Actors

Table 8.1 – Types of attacker objectives

Data exfiltration

Data exfiltration is a crucial step in the cyber kill chain where attackers transfer stolen data from the target network to their controlled environment. This phase is vital for attackers as it represents the completion of their efforts to infiltrate the system and gather sensitive information. Understanding the methodologies and techniques used in data exfiltration is essential for devising effective countermeasures.

Methods of data exfiltration

Data exfiltration techniques are a critical aspect of the cyber kill chain's *Exploitation* phase. Understanding these methods is essential for cybersecurity professionals to develop effective defense strategies. In this section, we will explore three primary methods used for data exfiltration:

- **Network-based exfiltration**: Cybercriminals often employ network-based data exfiltration techniques to siphon sensitive information from compromised systems. This involves the use of covert channels, such as **command and control (C2)** communication, which allows malicious actors to establish and maintain hidden communication channels within the compromised network. Detecting and countering these channels is a complex but crucial task in defending against data exfiltration.

- **Data compression and encryption**: Attackers commonly use data compression and encryption to mask stolen data during transmission. This approach helps them evade detection by making the data appear innocuous. Cybersecurity professionals must have a deep understanding of these techniques, as it enables them not only to recognize such activities but also to enhance data protection measures and encryption practices.

- **Covert channels**: Covert channels represent another sophisticated method used by attackers for data exfiltration. These channels allow them to move data discreetly, often using techniques such as steganography, where data is hidden within seemingly innocent files or images. Detecting covert channels requires a keen awareness of evolving exfiltration strategies and a commitment to innovation in detection methods.

Each of these methods demands a unique set of skills and tools for detection and mitigation. Cybersecurity experts need to stay updated on the latest developments in these areas to effectively combat data exfiltration.

Techniques used by attackers in data exfiltration

Attackers, showing their ingenuity, use sophisticated techniques to exfiltrate data from compromised networks. Each method is meticulously designed to evade detection by security systems, underscoring the need for constant vigilance and the importance of understanding these techniques to ensure the successful transfer of stolen data.

One standard method is direct transfer, which involves uploading stolen data to an external server using network protocols such as **File Transfer Protocol (FTP)**, **Hypertext Transfer Protocol (HTTP)**, or **Hypertext Transfer Protocol Secure (HTTPS)**. Attackers connect to an external server and transfer data through open network ports. Encrypted connections via HTTPS can obscure the transmitted data, making it harder to detect. This method is simple and efficient for transferring large volumes of data, with encryption helping to bypass some detection systems.

Email-based exfiltration, a prevalent and effective technique, is another method used by attackers. They send stolen data through email attachments or embed it within the email body, often leveraging legitimate email services to avoid raising suspicions. The data is usually compressed and/or encrypted before being sent to attacker-controlled email accounts. By using common email services such as Gmail or Outlook, attackers can often bypass security filters, making this method both easy to execute and highly effective. This prevalence underscores the urgency of addressing this common threat.

Attackers also favor uploading stolen data to cloud storage services such as Google Drive, Dropbox, or OneDrive. These services use encrypted connections, making the data transfer more secure and less likely to be detected. Attackers create accounts on cloud storage platforms and upload the data using secure connections (SSL/TLS), often leveraging the trusted nature of these services to avoid network security measures.

Steganography, the practice of hiding data within other files such as images, videos, or audio files, is another sophisticated method attackers use. Attracting data within the least significant bits of the host file allows attackers to transfer modified files without arousing suspicion. Special tools or scripts are then used to extract the hidden data. This technique is particularly stealthy and difficult to detect with conventional scanning tools, as the exfiltrated data appears benign.

DNS tunneling involves encapsulating data within DNS queries and responses. This method can evade many security controls since DNS traffic is typically allowed and trusted. Data is split into small chunks and encoded into DNS requests, then sent to a malicious DNS server controlled by the attacker. The server then decodes the data and assembles it. The small data chunks and the trusted nature of DNS traffic make this method highly effective and hard to detect.

Finally, physically removing data using USB drives or other removable media is a straightforward exfiltration method, often involving insider threats. Data is copied to USB drives or other removable storage devices and physically removed from the premises. Attackers may use encryption to protect the data during transport. This method bypasses network security measures and can quickly exfiltrate large amounts of data.

Understanding these techniques is just the first step. It's crucial to use this knowledge to devise robust security strategies to detect and mitigate data exfiltration attempts. By comprehensively analyzing and monitoring these methods, organizations can better protect sensitive information from falling into the wrong hands, emphasizing the importance of proactive security measures.

Impact of data exfiltration

Successful data exfiltration can have profound and lasting consequences for organizations, affecting their financial health, reputation, and legal standing. In this section, we will delve into the multifaceted impact of data breaches, covering the extensive financial ramifications, the substantial damage to an organization's reputation, the complex regulatory and legal consequences, and the pivotal role of **Data Loss Prevention (DLP)** in mitigating these severe consequences:

- **Financial consequences**: Data breaches can lead to significant financial losses for organizations. Beyond immediate expenses for breach recovery, these losses may include regulatory fines, legal fees associated with litigation, and costs incurred during breach response and recovery. Organizations must recognize the wide-reaching financial implications of data exfiltration, including potential revenue loss due to reputational damage. Allocating resources for both immediate recovery and long-term financial stability is paramount.

  The financial consequences can be staggering. The *2020 Cost of a Data Breach Report* by IBM Security and Ponemon Institute found that the average total cost of a data breach globally was $3.86 million. These costs encompass both direct expenses, such as incident response, and indirect costs, including lost business due to diminished trust.

- **Reputation damage**: One of the most visible and detrimental impacts of data breaches is the damage they inflict on an organization's reputation. Successful data breaches erode trust among customers, partners, and stakeholders. Rebuilding trust is a complex and time-consuming process that involves not only communication efforts but also substantial investments in security measures and practices.

  Reputation damage can have long-lasting effects. A study conducted by Edelman Trust Barometer revealed that 74% of customers are more likely to lose trust in a company that experiences a data breach. Restoring trust necessitates a well-executed public relations strategy, enhanced security protocols, and ongoing efforts to demonstrate a commitment to data protection. The financial implications extend beyond immediate losses, as an organization's reputation is intertwined with its long-term success.

- **Regulatory and legal ramifications**: Organizations that experience data breaches may face intricate regulatory penalties and legal actions, particularly if they fail to comply with data protection laws. The legal consequences can encompass fines, lawsuits, regulatory sanctions, and other punitive measures. Staying well-informed about these legal aspects is imperative for organizations as non-compliance can magnify the financial burden and tarnish their reputation further.

  For example, the European Union's **General Data Protection Regulation (GDPR)** provides a framework for hefty fines, which can amount to €20 million or 4% of the company's global annual turnover, whichever is higher, for severe violations of data protection rules. These legal ramifications underscore the importance of meticulous data protection practices.

- **DLP**: Implementing effective DLP measures is a critical strategy for mitigating the impact of data exfiltration. DLP strategies encompass data classification, access controls, and real-time monitoring to prevent unauthorized data transfers. A thorough understanding of DLP is essential for organizations seeking to safeguard their sensitive data effectively. Furthermore, a well-implemented DLP program can help organizations avoid or mitigate financial, reputational, and legal consequences.

By addressing these multifaceted aspects of data breaches, organizations can take proactive measures to protect their critical assets, minimize the financial burden, rebuild their reputation, and ensure compliance with data protection laws.

Mitigation strategies

Mitigating data exfiltration and defending against sophisticated attackers requires a multi-layered approach, combining technical controls, policy enforcement, and continuous monitoring. Here are some effective strategies for protecting against data exfiltration:

- **Network monitoring and traffic analysis**: Continuous monitoring of network traffic is essential for detecting unusual patterns that may indicate data exfiltration attempts. Implementing **Intrusion Detection Systems (IDSs)** and **Intrusion Prevention Systems (IPSs)** can help identify and block suspicious data flows. These systems can be configured to flag anomalies, such as unexpected data transfers, large volumes of outbound traffic, or connections to known malicious IP addresses.

- **Data encryption**: Data encryption is a fundamental defense against data exfiltration, acting as a robust last line of defense even if attackers manage to infiltrate a network and access sensitive data. Understanding encryption techniques and best practices is crucial for organizations seeking to safeguard their critical assets effectively. Robust encryption mechanisms rely on advanced cryptographic algorithms to encode data, making it indecipherable to unauthorized users. Data encryption requires encryption software, secure key management, and robust policies and procedures. Encryption extends beyond data at rest to encompass data in transit, including information transferred over networks. Effective data encryption is essential to data protection and privacy regulations worldwide, such as GDPR and HIPAA. Modern encryption methods, such as end-to-end encryption, secure data across various platforms from email communications to instant messaging apps. The successful implementation of encryption measures depends not only on the technology but also on educating employees on encryption best practices and promoting a culture of data security.

- **Access controls and authentication**: Implementing strict access controls and authentication mechanisms can limit unauthorized access to sensitive data. **Role-based access control (RBAC)** ensures that users have only the permissions necessary for their roles, reducing the risk of data exposure. **Multi-factor authentication (MFA)** adds an extra layer of security, making it more difficult for attackers to use compromised credentials to gain access.

- **DLP solutions**: DLP solutions prevent unauthorized data transfers by monitoring and controlling data flows within and outside the organization. DLP tools can identify and block attempts to transfer sensitive data via email, cloud storage, USB devices, and other channels. These solutions can also enforce policies regarding data handling and provide alerts for potential data exfiltration events.

- **Security Information and Event Management (SIEM)**: SIEM systems are pivotal in detecting and responding to data exfiltration attempts. These systems provide real-time visibility into network activities, enabling organizations to detect anomalies and respond promptly to security incidents. SIEM systems offer a centralized platform that aggregates and analyzes security data from various sources, including logs, network traffic, and system events. SIEM solutions can identify suspicious patterns and potential security threats by monitoring and correlating this data. Alerts and notifications are generated when anomalies or security breaches are detected, allowing security teams to respond promptly. Advanced SIEM systems incorporate machine learning and artificial intelligence to enhance threat detection and reduce false positives. They empower organizations to address potential data exfiltration attempts and other security incidents proactively. Implementing SIEM systems requires careful planning, including selecting appropriate solutions, configuration, and ongoing management. It is a technology deployment and a framework for security operations, integrating people, processes, and technology.

- **Endpoint security**: Deploying comprehensive endpoint security solutions helps protect devices from being used as vectors for data exfiltration. Endpoint security includes antivirus software, **endpoint detection and response** (EDR) tools, and regular patch management to address vulnerabilities. These tools can detect and neutralize malware, monitor suspicious activities, and update systems with the latest security patches.

- **Cloud Access Security Brokers (CASBs)**: CASBs provide visibility and control over data in cloud environments. They enforce security policies for cloud applications, monitor user activities, and detect anomalies that may indicate data exfiltration attempts. By applying consistent security controls across all cloud services, CASBs can also ensure compliance with data protection regulations.

- **User education and training**: Educating employees about the risks and signs of data exfiltration is critical to any defense strategy. Regular training sessions can help employees recognize phishing attempts, suspicious emails, and other social engineering tactics attackers use. Encouraging a security-aware culture ensures that employees are vigilant and proactive in reporting potential threats.

- **Physical security controls**: Physical security measures should be in place to protect against data exfiltration via removable media. This includes restricting access to sensitive areas, using security cameras, and implementing policies for using USB drives and other removable devices. Regular audits and inspections can help identify and mitigate physical security risks.

- **Incident response planning**: A well-defined incident response plan enables organizations to respond quickly and effectively to data exfiltration incidents. The plan should outline procedures for identifying, containing, and mitigating the impact of data breaches. Regular drills and updates to the incident response plan ensure that the organization is prepared to handle real-world threats.

By integrating these mitigation and defense strategies, organizations can build a robust security posture that protects against data exfiltration. Continuous improvement and adaptation to emerging threats are essential for maintaining the effectiveness of these measures.

Real-life incidents demonstrating the final stage

In this section, we will delve into real-world examples of cyberattacks where attackers successfully carried out their objectives during the final stages of the cyber kill chain. These incidents serve as valuable case studies, illustrating how attackers exploit vulnerabilities and the consequences of their actions.

Case study 1 – Sony Pictures hack (2014)

In 2014, Sony Pictures Entertainment found itself at the center of one of the most high-profile and politically charged cyberattacks in recent history. The breach, attributed to North Korea, sent shockwaves through the entertainment industry and the cybersecurity world, underlining the far-reaching consequences of politically motivated cyberattacks. This incident serves as a vivid case study of the exploitation phase of the cyber cill chain, where the attackers successfully achieved their objectives. In this detailed analysis, we delve into the intricacies of the Sony Pictures hack, examining the objectives, tactics, and repercussions of this notorious breach.

The Sony Pictures hack was not a run-of-the-mill cyberattack. It was intrinsically political and driven by a specific motive. At its core, the objective of the attackers was twofold.

The immediate trigger for the cyberattack was the impending release of the comedy film *The Interview*, a satirical take on the North Korean regime. North Korean leaders viewed the movie as a direct affront to their regime, and they aimed to prevent its release. The attackers sought to create chaos and intimidate Sony into shelving the movie, ultimately disrupting its release.

In addition to disrupting the movie's release, the attackers had a secondary motive—stealing sensitive corporate data. The breach resulted in the exposure of a vast trove of corporate documents, emails, and other confidential information, which was subsequently leaked online. The stolen data included everything from financial records to personal employee information.

The Sony Pictures hack was executed with a high degree of sophistication. The attackers employed several tactics to achieve their objectives.

The initial entry point for the attackers was spearphishing emails. They used a variety of tactics to trick Sony employees into revealing their credentials and granting access to the corporate network. These phishing emails were designed to appear as if they came from legitimate sources, luring victims into unwittingly providing their login information.

Once inside Sony's network, the attackers deployed destructive malware known as *Destover*. This malware was specifically designed to wipe data from infected systems. In this phase of the attack, the objectives were clear—disrupting Sony's operations and erasing data. Computers at Sony Pictures were rendered inoperable, and sensitive data was deleted, causing significant damage.

Alongside the destructive aspect of the attack, the attackers also conducted data exfiltration. They carefully selected and exfiltrated sensitive documents, which were later released to the public or held as leverage against Sony. This data theft extended the impact of the attack beyond mere disruption.

The Sony Pictures hack had profound consequences. Sony Pictures suffered significant reputation damage, not only due to the disruption of the release of *The Interview* but also because of the exposure of sensitive corporate data. The leaked information included private emails and documents that could be embarrassing or damaging to the company.

The hack created geopolitical tensions between the U.S. and North Korea. The U.S. government attributed the attack to North Korea, and this attribution led to sanctions and diplomatic tensions. The hack became a subject of international scrutiny and debate.

The U.S. Department of Justice unsealed charges against a North Korean national in 2018 for their role in the Sony Pictures hack. While it's unlikely that the individual will ever face trial in the U.S., this action underscores the seriousness with which the U.S. government treated the attack.

The attribution of the attack to North Korea was a complex process involving a combination of technical, geopolitical, and intelligence analysis. The attackers left behind distinctive digital fingerprints and utilized infrastructure linked to previous North Korean campaigns, providing crucial clues to investigators.

The Sony Pictures hack stands as a stark reminder of the far-reaching consequences of politically motivated cyberattacks. It underscores several critical lessons.

Despite advancements in cybersecurity, spearphishing remains a highly effective tactic for initiating cyberattacks. Employee education and strong email security measures are essential defenses.

While data theft is often the primary focus of discussions surrounding cyberattacks, the deployment of destructive malware with the intent to disrupt and damage systems is a real threat that organizations must consider.

Politically motivated cyberattacks have broader geopolitical implications. Attribution and response to such attacks can significantly impact international relations and diplomatic efforts.

The exposure of sensitive corporate data underlines the importance of robust data security measures. Organizations must protect their data not only from theft but also from destruction.

Case study 2 – SWIFT banking heists (2016 and 2018)

Some instances stand out not only for their audacity but also for the substantial financial implications they have. The SWIFT banking heists of 2016 and 2018 are notable examples, demonstrating how cybercriminals targeted global financial systems with the objective of stealing significant funds through fraudulent transactions. These incidents not only exposed the vulnerabilities within the SWIFT international banking network but also underscored the need for robust security measures and increased vigilance in the financial sector.

The **Society for Worldwide Interbank Financial Telecommunication** (**SWIFT**) is a global messaging network used by financial institutions to securely exchange information and execute transactions. SWIFT is a critical component of the global financial infrastructure, providing a standardized and secure means of conducting financial transactions, including the transfer of funds between banks across borders.

2016 – Bangladesh Bank heist

In February 2016, cybercriminals executed a meticulously planned heist targeting Bangladesh Bank. The attackers utilized fraudulent SWIFT messages to transfer $81 million from the bank's account with the Federal Reserve Bank of New York to accounts in the Philippines. The attackers sought to launder the stolen funds through casinos in the Philippines.

The attack was notable for several reasons. The attackers demonstrated a deep understanding of the SWIFT system and exploited vulnerabilities in both the bank's security and the SWIFT software to execute the fraudulent transactions.

The scale of the heist and the target, the central bank of a nation, sent shockwaves through the global financial community.

The incident revealed significant security lapses within Bangladesh Bank's systems, including the absence of firewalls and an ineffective intrusion detection system.

2018 – The Ecuadorian cyber heist

In 2018, a group of cybercriminals executed a heist targeting the central bank of Ecuador. The attackers transferred $12 million from the bank's accounts to foreign accounts. Similar to the Bangladesh Bank heist, fraudulent SWIFT messages were used to facilitate the transfers.

The attackers in this incident employed tactics reminiscent of the Bangladesh Bank heist, demonstrating that the same vulnerabilities in the SWIFT system and the global banking network were still exploitable. This heist highlighted the international nature of cyberattacks, as funds were transferred across borders.

Recovering the stolen funds in these cases is a complex and protracted process, and the success of such efforts is far from guaranteed.

The SWIFT network and cybersecurity challenges

The SWIFT network is a prime target for cybercriminals due to the significant financial transactions it facilitates. The SWIFT system relies on standardized message formats, and these messages contain sensitive financial information. Cybercriminals, in both the Bangladesh Bank and Ecuadorian heists, exploited vulnerabilities at multiple points in the transaction process, often starting with the compromise of the target bank's internal systems.

The SWIFT network faced several cybersecurity challenges in these heists.

The attackers often initiated their operations by compromising the security of the user environments in the targeted banks. These compromises allowed them to generate fraudulent SWIFT messages that appeared legitimate.

In both cases, the attackers gained access to the SWIFT credentials of the targeted banks. This level of access enabled them to initiate transactions and even override security controls.

Weaknesses in MFA and dual authorization processes within the SWIFT system allowed the attackers to bypass these safeguards.

The SWIFT banking heists serve as stark reminders of the ever-present threat of cyberattacks in the financial sector. Lessons learned from these incidents have led to the implementation of numerous security measures to safeguard the SWIFT network and global financial systems.

SWIFT introduced the **Customer Security Program (CSP)** to bolster security within its network. The CSP includes mandatory security controls and guidelines for member institutions to follow.

Financial institutions now place a stronger emphasis on monitoring and detecting unusual or suspicious transaction patterns, enabling the early detection of potential cyberattacks.

Enhancements in MFA and dual authorization processes within the SWIFT network make it more challenging for attackers to gain access to critical systems.

The global nature of cyberattacks necessitates international cooperation and information sharing to combat threats effectively.

The SWIFT banking heists represent a significant milestone in the ongoing struggle to secure global financial systems against cyber threats. They underscore the need for constant vigilance, robust security measures, and international collaboration to safeguard the integrity of financial transactions in an interconnected world.

Proactive defensive strategies to thwart objectives-based attacks

Objectives-based attacks are sophisticated and targeted, often requiring a robust defensive strategy to prevent or mitigate their impact. In this section, we'll explore key defensive strategies, including DLP, network segmentation, and incident response, as a means to thwart objectives-based attacks.

DLP

DLP stands for data loss prevention. Let's go through what it entails:

- **Understanding data exfiltration**: Objectives-based attacks often involve the theft or exfiltration of sensitive data, making a robust understanding of data exfiltration methods paramount for effective DLP implementation. Cybersecurity professionals need to be well-versed in the various techniques employed by malicious actors to siphon data. This includes methods such as covert channels, network-based exfiltration, and data compression. By comprehending the intricacies of these tactics, organizations can better design their DLP strategies to counteract them effectively.

- **Content inspection and policy enforcement**: DLP solutions employ advanced content inspection and policy enforcement mechanisms to monitor and control data movements within an organization's network. These systems continuously scan data in transit and at rest, ensuring that it adheres to established security policies and protocols. When violations are detected, DLP solutions can take immediate actions, such as blocking the data transfer, alerting administrators, or applying encryption to protect the data. A comprehensive understanding of how content inspection and policy enforcement function within DLP solutions is vital for organizations to maintain granular control over their data and enforce their security policies consistently.

- **User education**: While the technical aspects of DLP are crucial, it's essential not to overlook the human factor. User awareness and education play integral roles in the success of DLP strategies. Educating employees about the significance of data protection and the far-reaching consequences of data breaches is a fundamental component of DLP. When employees are well-informed about security best practices and the importance of adhering to data protection policies, they become active participants in safeguarding sensitive information. User education can substantially reduce the risk of data leakage resulting from human error, inadvertently strengthening the organization's overall security posture.

By integrating these mitigation and defense strategies, organizations can build a robust security posture that protects against data exfiltration. Continuous improvement and adaptation to emerging threats are essential for maintaining the effectiveness of these measures.

Implementing a Zero-Trust architecture

Zero-Trust Architecture (**ZTA**) represents a paradigm shift in network security. It advocates for a model where no user or system is trusted by default, irrespective of their location within the network. This approach aligns seamlessly with network segmentation, further strengthening an organization's security posture by enforcing strict access controls, verification, and continuous monitoring at every level of network access.

Core principles of ZTA

ZTA is built on core principles that redefine network security by ensuring that no entity, whether user or system, is trusted by default. These principles form the foundation of a zero-trust strategy and guide the implementation of security measures designed to protect organizational assets from internal and external threats:

- **Never trust, always verify**: The cornerstone of ZTA is the principle of *never trust, always verify*. This means no user or device is trusted by default, whether inside or outside the network. Every access request must be authenticated and authorized before access is granted. This principle recognizes that threats can exist within and outside the network perimeter, and continuous verification is necessary to maintain security.

- **Least privilege access**: ZTA enforces the principle of least privilege access, which ensures that users and systems are granted the minimum level of access necessary to perform their functions. By limiting access rights to only what is essential, the potential damage from compromised accounts or devices is minimized. This principle reduces the attack surface and helps prevent attackers' lateral movement within the network.

- **Micro-segmentation**: Micro-segmentation involves dividing the network into smaller, isolated segments or zones, each with access controls and security policies. This granular approach to network segmentation helps contain breaches and limits the spread of attacks. Even if an attacker gains access to one segment, they will face additional barriers when attempting to move laterally within the network.

- **Dynamic policies**: These are essential for maintaining robust security in an ever-evolving threat landscape. These policies adapt in real time to changes in the network environment, user behavior, and threat intelligence. Unlike static security measures, dynamic policies continuously evaluate and enforce access controls based on a comprehensive set of contextual data, such as user identity, device health, location, and the sensitivity of the requested resources. This real-time adaptability ensures that security decisions are always relevant and practical, minimizing the risk of unauthorized access and swiftly mitigating potential threats. By leveraging automation and advanced analytics, dynamic policies enhance the agility and resilience of the security posture, making it difficult for attackers to exploit vulnerabilities and maintain persistent access within the network.

- **Continuous monitoring and inspection**: Continuous monitoring and real-time inspection of network traffic and user activity are critical components of ZTA. This principle involves using advanced analytics, machine learning, and threat intelligence to detect and respond to anomalies and potential security incidents. Organizations can identify and mitigate threats by continuously analyzing behavior and traffic patterns before they cause significant damage.

- **Secure access for all resources**: ZTA mandates secure access to all resources, whether on-premises or in the cloud. This principle ensures that every access request, regardless of the resource's location, undergoes the same rigorous authentication and authorization process. It extends security measures uniformly across the entire IT environment, including remote and mobile devices.

- **Identity and Access Management (IAM)**: Effective IAM is essential in ZTA. This principle involves verifying the identity of users and devices and ensuring they have the appropriate permissions to access specific resources. MFA and strong password policies are critical components of IAM, providing additional layers of security beyond simple username and password combinations.

- **Data protection**: Protecting data at rest and in transit is a fundamental principle of ZTA. Data encryption ensures that it remains unreadable and secure even if data is intercepted or accessed without authorization. This principle applies to all types of data, including sensitive information, intellectual property, and personal data, ensuring compliance with data protection regulations and standards.

- **Policy enforcement**: ZTA relies on consistently enforcing security policies across the entire network. These policies dictate who can access resources, under what conditions, and for what purposes. Automated policy enforcement mechanisms help ensure that security controls are applied uniformly, reducing the risk of human error and ensuring compliance with organizational security standards.

- **Assume breach**: The *assume breach* principle underpins the ZTA approach, acknowledging that no security measure is foolproof. Organizations should operate assuming that a breach has occurred or will occur and design their security architecture accordingly. This proactive mindset drives the implementation of robust detection, response, and recovery mechanisms to minimize the impact of breaches.

By adhering to these core principles, organizations can build a robust ZTA that mitigates risks, protects sensitive data, and ensures secure access to resources. ZTa represents a proactive and comprehensive approach to cybersecurity designed to address the evolving threat landscape and safeguard organizational assets in an increasingly complex digital environment.

Key technologies for implementing Zero Trust

Here are some key technologies used to achieve Zero Trust:

- **IAM**: IAM systems are crucial for Zero Trust. They provide tools for managing user identities, roles, and access permissions. These systems ensure that only authenticated and authorized users can access specific resources based on their roles and responsibilities.

- **MFA**: MFA adds an extra layer of security by requiring users to provide multiple verification forms before granting access. This could include something the user knows (password), something the user has (security token), and something the user is (biometric verification).

- **Continuous monitoring and analytics**: Zero Trust constantly monitors network traffic and user activities to detect and respond to threats in real time. Advanced analytics and machine learning algorithms can identify unusual patterns and behaviors indicative of a potential breach.

- **Network Access Control (NAC)**: NAC solutions enforce security policies at the network entry points, ensuring that only compliant and authenticated devices can connect. These solutions can dynamically adjust access permissions based on the device's security posture and the user's role.

- **Data encryption**: Encrypting data at rest and in transit is essential in a zero-trust model. This ensures that even if data is intercepted or accessed without authorization, it remains unreadable and secure.

- **Endpoint security**: Comprehensive endpoint security measures protect devices from being compromised and used as entry points for attackers. These measures include antivirus software, EDR tools, and regular security updates.

- **Software-Defined Perimeter (SDP)**: Implementing micro-segmentation divides the network into isolated segments, each protected with its security controls. An SDP creates a virtual boundary around resources, allowing access only to authenticated and authorized users, effectively hiding resources from unauthorized users.

Implementing Zero-Trust principles

Implementing Zero-Trust principles requires a systematic approach to enhance security measures across an organization's network. This process involves several critical steps that ensure robust security controls and continuous monitoring:

1. **Assessment and planning**: Begin by thoroughly assessing the existing network infrastructure, and identifying critical assets, data flows, and potential vulnerabilities. Then, develop a comprehensive plan that outlines the steps and technologies required to transition to a Zero-Trust model.

2. **Phased implementation**: Transitioning to Zero Trust should be a phased process, starting with the most critical areas. Implement IAM and MFA solutions, followed by network segmentation and continuous monitoring. Gradually extend Zero-Trust principles to all parts of the network.

3. **Employee training and awareness**: It is crucial to educate employees about Zero-Trust principles and the importance of security practices. Regular training sessions and awareness programs help foster a security-first culture within the organization.

4. **Continuous improvement**: Zero Trust is not a one-time implementation but a constant process. Regularly review and update security policies, conduct vulnerability assessments, and refine access controls to adapt to evolving threats and changes in the network environment.

By embracing ZTA, organizations can significantly enhance their security posture, ensuring that every access request is thoroughly and continuously monitored. This approach minimizes the risk of unauthorized access and data breaches, creating a resilient defense against modern cyber threats.

Network segmentation

Network segmentation is a strategic approach to enhance the security of a network by dividing it into smaller, isolated segments or zones. This strategy plays a pivotal role in building a resilient defense against objectives-based attacks, making it more challenging for attackers to achieve their objectives and limiting lateral movement within the network. Here are some key strategies to consider:

-   **Segmenting the network**: Network segmentation involves the creation of security zones within a network. Each zone has its access controls and security policies, ensuring that even if one segment is compromised, the attacker's ability to move laterally to other segments is constrained. This is particularly significant in preventing attackers from easily moving from, for example, an initial point of compromise to critical data repositories or sensitive systems. Understanding the principles of network segmentation is crucial. It requires careful planning, including identifying which segments are most sensitive and need the highest level of protection. Proper segmentation can be achieved through firewalls, **virtual LANs (VLANs)**, and access controls.

-   **Micro-segmentation**: Micro-segmentation takes network segmentation a step further by dividing segments into smaller, granular units. Each unit corresponds to individual applications, workloads, or even specific assets, which are then isolated from each other. This fine-grained control enhances security by reducing the attack surface and minimizing the lateral movement an attacker can undertake within the network.

Understanding the intricacies of micro-segmentation empowers organizations to tailor access controls to the specific needs of their applications and data. For example, high-value databases may have the most stringent access controls, while less critical systems have more permissive settings. By segmenting the network into these granular units, organizations achieve a higher level of control and security.

Network segmentation, when combined with ZTA and micro-segmentation, provides a robust defense against objectives-based attacks. These strategies limit lateral movement, minimize the attack surface, and ensure that trust is never assumed, ultimately enhancing an organization's security posture and resilience against targeted cyber threats.

Responding to a data exfiltration attack

Responding to a data exfiltration attack involves a well-coordinated set of actions to contain the breach, mitigate damage, and prevent future incidents. This section outlines a comprehensive response strategy, divided into critical phases: detection and identification, containment, eradication, recovery, notification and reporting, post-incident analysis, and improving security posture.

Incident response

Incident response is critical to an organization's cybersecurity strategy, especially when dealing with objectives-based attacks. A well-prepared and structured incident response plan is essential to mitigate and recover from security incidents effectively:

- **Developing an incident response plan**: Organizations must establish a comprehensive incident response plan that outlines procedures to follow when detecting objectives-based attacks. An effective plan involves a step-by-step response process, the assignment of responsibilities to incident response teams, and communication strategies for internal and external stakeholders. The incident response plan should define the roles and responsibilities of key personnel, such as incident managers, forensic analysts, and legal representatives. It must specify clear guidelines for assessing the severity of incidents, reporting procedures, and communication channels with internal teams, executives, and relevant authorities. A well-designed plan also incorporates legal and regulatory considerations, ensuring the organization complies with data breach notification requirements.

- **Timely detection and containment**: Rapid detection and containment are paramount in the face of objectives-based attacks. Incident response teams should be well-trained to promptly identify, assess, and mitigate these attacks. This involves leveraging advanced detection technologies, real-time monitoring, and automated responses to stop the attack's progression and minimize damage. Early detection is essential to reduce the attacker's dwell time within the network, limiting their ability to achieve their objectives. Detection mechanisms include IDSs, SIEM solutions, and anomaly detection tools. Automated responses can range from isolating affected systems to blocking malicious traffic. Timely containment measures are crucial in minimizing the impact of the attack and preventing its spread to other network segments.

- **Post-incident analysis**: After an incident has been effectively addressed, a thorough post-incident analysis is essential to understand the attack's scope and refine defensive strategies. Lessons from each incident should be incorporated into the incident response plan, security policies, and ongoing security measures. Continuous improvement is critical in adapting to evolving threats. Post-incident analysis involves a detailed examination of the attack's **tactics, techniques, and procedures** (**TTPs**). It aims to identify vulnerabilities or weaknesses in the organization's defense that the attacker exploited. This knowledge informs security enhancements, such as updates to intrusion detection rules, security awareness training improvements, and incident response procedure adjustments. Additionally, post-incident analysis can provide valuable insights for threat intelligence, helping organizations better prepare for future threats.

Detection and identification

In this phase, detecting and identifying the data exfiltration activity promptly is crucial. Immediate actions and the deployment of appropriate tools and techniques can help identify the incident's nature and scope.

Immediate actions include the following:

- **Alert review**: Verify the alerts from monitoring tools such as IDSs, IPSs, SIEM, and DLP systems to confirm the exfiltration activity.

- **Incident identification**: Identify the scope and nature of the data exfiltration incident, including affected systems, data types, and the volume of exfiltrated data.

Tools and techniques include the following:

- Utilize SIEM systems to correlate and analyze logs

- Conduct network traffic analysis to identify suspicious data transfers

- Deploy EDR tools to check for malware and unauthorized activities

Containment

Once a data exfiltration incident is identified, swift containment measures are essential to prevent further data loss and limit the attack's impact. The following actions and tools are recommended for effective containment.

Immediate actions include the following:

- **Network segmentation**: Isolate affected systems from the rest of the network to prevent further data loss.

- **Access revocation**: Disable compromised accounts and change credentials to block unauthorized access.

- **Quarantine infected devices**: Disconnect devices suspected of being compromised to stop the exfiltration.

Tools and techniques include the following:

- Implement firewall rules to block outbound traffic to known malicious IP addresses

- Use NAC to enforce isolation policies

- Employ endpoint security solutions to quarantine infected devices

Eradication

After containing the incident, the next step is to eradicate the threat from the network. This involves removing any malware, patching vulnerabilities, and shutting down the attack vectors used by the attackers.

Immediate actions include the following:

- **Malware removal**: Use antivirus and anti-malware tools to clean infected systems.
- **Patch vulnerabilities**: Apply security patches to fix exploited vulnerabilities.
- **Close attack vectors**: Identify and shut down methods attackers use to gain entry, such as unpatched software or weak passwords.

The tools and techniques to eradicate threats include the following:

- Deploy patch management systems to ensure timely updates
- Use malware analysis tools to understand the attacker's tools and techniques
- Conduct vulnerability scans to identify and rectify security gaps

Recovery

Following eradication, the recovery phase focuses on restoring systems and data integrity. This ensures that all systems are clean, and operations can resume securely.

Immediate actions include the following:

- **System restoration**: Restore affected systems from clean backups to ensure they are malware-free.
- **Data integrity check**: Verify the integrity of critical data and ensure no unauthorized changes have been made.
- **Monitor systems**: Implement enhanced monitoring on restored systems to detect any signs of residual malicious activity.

The tools and techniques for recovery include the following:

- Leverage backup and disaster recovery solutions for system restoration
- Use data integrity verification tools to ensure data accuracy
- Continuously monitor systems using SIEM and EDR systems

Notification and reporting

Timely and accurate notification and reporting are critical components of incident response. This involves communicating with internal stakeholders, regulatory bodies, and affected customers to manage the incident's impact effectively.

Immediate actions include the following:

- **Internal communication**: Inform internal stakeholders, including management and IT teams, about the breach and the actions taken to respond.

- **Legal and compliance reporting**: Report the incident to relevant regulatory bodies as required by GDPR, HIPAA, or state data breach notification laws.

- **Customer notification**: If sensitive customer data is involved, notify affected customers and provide guidance on protective measures they can take.

The tools and techniques for notification and reporting include the following:

- Utilize incident response communication plans for internal notifications
- Follow legal and compliance frameworks to ensure regulatory adherence
- Implement public relations and customer support strategies for effective communication

Improving security posture

Defending against objectives-based attacks requires a multifaceted approach, combining DLP, network segmentation, and incident response strategies. Organizations can reduce the likelihood and impact of such targeted cyberattacks by implementing these defensive measures. Continuous improvement and adaptation to emerging threats are essential for maintaining the effectiveness of these measures.

Immediate actions include the following:

- **Policy review**: Update security policies and procedures based on the findings from the incident analysis.

- **Employee training**: Conduct training sessions to raise awareness about security practices and the importance of vigilance.

- **Technological enhancements**: Implement advanced security technologies, such as improved encryption methods, enhanced DLP solutions, and AI-driven threat detection systems.

The tools and techniques to improve security posture include the following:

- Utilize policy management tools to streamline updates
- Launch security training programs and awareness campaigns
- Deploy advanced security solutions such as machine learning-based anomaly detection and behavior analysis tools

By following these steps, organizations can effectively respond to a data exfiltration attack, mitigate damage, and strengthen their defenses against future threats. The goal is to minimize the breach's impact, restore normal operations, and enhance overall security posture to prevent recurrence.

The eighth phase of the Cyber Kill Chain

The eighth step of the cyber kill chain is often called **monetization**, but it's essential to recognize that attackers' objectives can vary widely beyond mere financial gain. Depending on the motive, this phase could involve espionage, data exfiltration for strategic advantages, or releasing sensitive documents to harm reputations or disrupt operations. Attackers may achieve their goals without publicly disclosing stolen data, especially in cases where the value lies in the strategic use or manipulation of the information rather than selling it on illicit markets.

Preventing attacks from progressing to this final stage, regardless of the attacker's ultimate objective, is crucial for defenders. Effective mitigation, such as early threat detection and robust incident response strategies, helps to disrupt the attack chain before the attacker can fully exploit their access. For instance, techniques such as network segmentation, data encryption, and stringent access controls can impede attackers' ability to exfiltrate data or establish control over critical systems, thereby diminishing the attack's value. Even when data is never publicly released, these defensive measures retain their importance by reducing the attacker's success in achieving their intended payout, whether monetary, strategic, or reputational.

Organizations should emphasize detecting and neutralizing threats early and implementing controls that can limit the damage and reduce the target's attractiveness. This will mitigate the impact of cyberattacks and preserve the integrity of their operations.

Summary

In the final stage of the cyber kill chain, known as *Actions on Objectives*, attackers seek to accomplish their primary goals, ranging from data theft to system sabotage. This chapter delved into the intricacies of this stage, highlighting various attack objectives, profiling attackers, exploring the methods and impacts of data exfiltration, and outlining comprehensive mitigation and response strategies. Attackers have diverse objectives, including data theft, disruption, espionage, financial gain, and destruction. Understanding attacker profiles, such as hacktivists, cybercriminals, nation-state actors, and insiders, is essential for anticipating and defending against these threats.

Data exfiltration, a critical component of many attacks, involves transferring stolen data from the target network to an attacker-controlled environment using direct transfer, email, cloud storage, steganography, DNS tunneling, and USB devices. The impact of data exfiltration is not just significant; it can be devastating, leading to financial loss, reputational damage, operational disruption, and legal repercussions. To mitigate these risks, organizations must urgently implement robust defense strategies, including continuous monitoring, data encryption, access controls, and endpoint security.

Real-life incidents demonstrate the severe consequences of successful data exfiltration, emphasizing the importance of proactive defensive strategies. Organizations should adopt a multi-faceted approach, integrating ZTA, network segmentation, and incident response plans to thwart objectives-based attacks. In a data exfiltration attack, a structured reaction involving detection, containment, eradication, recovery, and post-incident analysis is crucial for minimizing damage and strengthening future defenses. By understanding the final stage of the cyber kill chain and implementing comprehensive security measures, organizations can better protect themselves against sophisticated cyber threats. In the next chapter, we will delve into advanced technologies like machine learning and artificial intelligence and the future of the cyber kill chain.

Further reading

Following are some bonus reading materials for you to check out:

- *Comprehensive overview of cyber kill chains, their components, and strategies for implementation*: https://www.splunk.com/en_us/blog/learn/cyber-kill-chains.html

- *Original Lockheed Martin Cyber Kill Chain® framework explanation and resources*: https://www.lockheedmartin.com/en-us/capabilities/cyber/cyber-kill-chain.html

- *Detailed explanation of the Cyber Kill Chain, its stages, and comparison with other frameworks*: https://www.proofpoint.com/au/threat-reference/cyber-kill-chain

Get This Book's PDF Version and Exclusive Extras

UNLOCK NOW

Scan the QR code (or go to packtpub.com/unlock). Search for this book by name, confirm the edition, and then follow the steps on the page.

Note: Keep your invoice handly. Purchase made directly from Packt don't require one.

9

Cyber Security Kill Chain and Emerging Technologies

Cybersecurity threats are becoming increasingly sophisticated, rendering traditional defense mechanisms insufficient in combating modern adversaries. Attackers now employ advanced tactics that leverage automation, real-time exploitation, and stealth to circumvent static security measures. To address these challenges, organizations must embrace innovative technologies that redefine the **Cyber Kill Chain (CKC)** and bolster their security frameworks. This chapter explores a range of transformative technologies, including **Artificial Intelligence (AI)**, blockchain, deception technologies, edge computing, synthetic data and **Privacy-Enhancing Technologies (PETs)**, and **Post-Quantum Cryptography (PQC)**. These cutting-edge advancements provide defenders with the ability to respond more intelligently, efficiently, and effectively across every phase of the CKC.

AI leads the charge by automating threat detection, enhancing strategic defenses, and countering human attackers with unmatched speed and precision. Blockchain strengthens data security and transparency with its decentralized, tamper-resistant architecture, ensuring secure communications throughout the kill chain. Deception technologies play a pivotal role by diverting attackers into controlled environments, allowing defenders to observe, analyze, and neutralize threats without jeopardizing critical systems. Similarly, edge computing enables faster, real-time threat detection and response by processing data closer to its source, minimizing latency and improving network efficiency.

Emerging technologies such as synthetic data and PETs are increasingly vital in cybersecurity. Synthetic data mimics real-world datasets without exposing sensitive information, providing a secure way to train AI models and detect advanced threats. PETs, including techniques such as differential privacy and homomorphic encryption, safeguard user data while enhancing threat analysis, making them indispensable in a privacy-conscious landscape. Meanwhile, PQC addresses the looming threat of quantum computing, offering encryption algorithms resistant to quantum-based attacks and ensuring long-term data security in the face of future technological advancements.

This chapter not only examines the current applications of these technologies but also highlights their evolving roles in reshaping the cybersecurity landscape. By adopting these advancements, organizations can develop adaptive and proactive defense strategies to address modern attack vectors effectively. The discussion sets a foundation for further exploration into these tools, equipping practitioners with the knowledge to strengthen their security postures while navigating emerging opportunities and challenges.

We will cover the following topics in the chapter:

- The role of AI in the CKC

- Blockchain in the CKC

- Deception technologies in disrupting cyberattacks

- Edge computing and 5G in enhancing real-time security

- Synthetic data and PETs

- Post-quantum cryptography

AI in the CKC

Did you know that AI-driven cyberattacks are increasing in complexity and sophistication every year? As attackers evolve their tactics, traditional defenses often fall short, requiring organizations to adopt advanced solutions. AI is at the forefront of this transformation, offering capabilities that redefine how cybersecurity teams detect, respond to, and neutralize threats. AI's ability to process vast amounts of data, predict malicious activity, and adapt in real time makes it a game-changer across all phases of the CKC. AI offers better security without a big budget. It watches millions of activities every second. It finds threats faster than human teams. A 2023 Cisco study found that AI security lowers the time to spot threats from 60 days (about 2 months) to 7 hours. 41% of small financial firms already use blockchain to secure data. A 2024 Deloitte report said blockchain stops data tampering by keeping records permanent and verifiable.

The following list outlines how AI can be strategically implemented across the different phases of the CKC to enhance threat detection, response, and mitigation:

- **Reconnaissance**: AI systems excel at monitoring extensive datasets to detect early signs of potential threats. Machine learning models can identify anomalies in network traffic and flag reconnaissance activities, enabling proactive measures against potential attacks.

- **Weaponization**: AI-driven threat analysis systems can dissect malware, anticipate attack vectors, and simulate scenarios to predict malicious behavior. This capability enables organizations to develop preemptive defense mechanisms.

- **Delivery**: In the *Delivery* phase, AI enhances defenses against phishing and malware distribution. **Natural Language Processing (NLP)** tools can detect sophisticated phishing attempts, while real-time URL analysis prevents access to harmful websites.

- **Exploitation**: AI accelerates vulnerability management by automating the identification of critical weaknesses and prioritizing remediation. It can detect zero-day exploits and predict future vulnerabilities, helping organizations mitigate risks proactively.

- **Installation**: AI strengthens endpoint security by identifying abnormal behaviors during software installation, such as unauthorized file executions or registry modifications. Automated response systems can quarantine affected endpoints, neutralizing threats before they spread.

- **Command and Control (C2)**: AI disrupts attacker communication channels by analyzing traffic for irregularities and blocking suspicious outbound connections. Enhanced by deception technologies, AI-powered honeypots can lure attackers and reveal their tactics.

- **Actions on objectives**: AI safeguards sensitive data by monitoring for unauthorized transfers and flagging anomalies, such as unusual data flows or large file movements. Automated recovery and forensic analysis streamline incident response and minimize damage during breaches.

Types of AI-based cyberattacks

AI has redefined the tactics of cyber attackers, enabling highly sophisticated, adaptive, and scalable threats. From AI-powered phishing to deepfakes and advanced malware, attackers now exploit AI to bypass defenses and deceive targets with unprecedented precision. This section examines key types of AI-based cyberattacks and their real-world implications.

AI-enhanced phishing and social engineering

AI-powered phishing attacks have become increasingly sophisticated, leveraging natural language processing to create highly convincing and personalized messages. Indicators that previously alerted users to phishing emails, such as typos and odd language patterns due to poor writing skills by malicious actors, are now significantly less common due to advancements in AI-generated content. For instance, in 2023, attackers used AI to craft highly convincing SMS phishing messages in a campaign against Activision, successfully tricking employees into revealing sensitive information and granting access to the company's employee database. This evolution underscores the necessity for more advanced, AI-driven defensive systems that can detect subtle contextual and behavioral anomalies, moving beyond traditional detection methods.

The following screenshot illustrates an AI-generated phishing email that mimics legitimate communications, highlighting the difficulty in identifying malicious content:

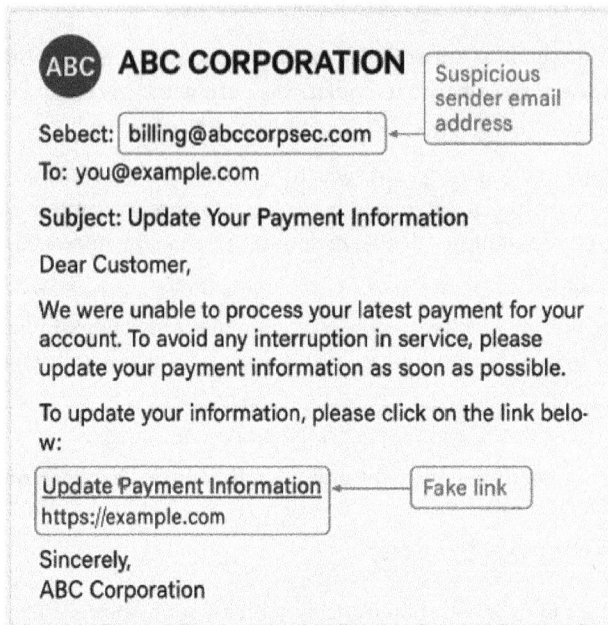

ABC CORPORATION

Suspicious sender email address

Sebect: billing@abccorpsec.com

To: you@example.com

Subject: Update Your Payment Information

Dear Customer,

We were unable to process your latest payment for your account. To avoid any interruption in service, please update your payment information as soon as possible.

To update your information, please click on the link below:

Update Payment Information
https://example.com

Fake link

Sincerely,
ABC Corporation

Figure 9.1 – An example of an AI-crafted phishing email is indistinguishable from authentic messages

Deepfakes and synthetic media

Deepfakes and synthetic media represent a concerning type of AI-generated attack, using advanced algorithms to create or manipulate audio, video, or images for malicious purposes. A striking example occurred in 2024 when fraudsters used deepfake technology to impersonate a company's chief financial officer during a video conference, resulting in a $25 million loss for a Hong Kong-based company. This case demonstrates the potential for AI-generated deepfakes to bypass traditional security measures and human intuition, posing significant risks to organizations. Robust forensic detection and investigation methods using machine learning, signal processing, traditional forensics, and file analysis are essential to combat such threats.

AI-enhanced malware and ransomware

AI-powered malware and ransomware have evolved to become more adaptive and evasive. These malicious programs use machine learning algorithms to modify their behavior in real time, evading detection by traditional security solutions. In 2024, the Dark Angels ransomware gang employed AI-powered ransomware that could dynamically mutate its code to avoid antivirus detection, resulting in a successful extortion of $75 million from a victim. Such incidents underscore the growing threat of AI-driven malware that can learn and adapt to its environment.

Adversarial AI attacks

Adversarial AI attacks aim to manipulate AI systems by introducing carefully crafted inputs that cause misclassification or erroneous outputs. While specific real-world examples of successful adversarial attacks against major AI systems are not widely publicized, researchers have demonstrated their potential impact. For instance, engineers at Duke University hacked the radar systems of autonomous vehicles, making them hallucinate other cars, which could potentially cause major accidents in real-world scenarios.

Evasion attacks

Evasion attacks leverage AI to help malicious actors avoid detection by security systems. A real-world example occurred in 2018 when researchers at Tencent Keen Security Lab demonstrated an evasion attack against Tesla's Autopilot system. They placed small, innocuous-looking stickers on the road surface in a specific pattern, confusing the AI system and causing it to identify a non-existent lane and follow an incorrect path. This example highlights how small, targeted changes in input data can manipulate AI systems, potentially leading to dangerous consequences.

Automated reconnaissance and targeting

AI has significantly enhanced automated reconnaissance and targeting, allowing attackers to more efficiently identify and exploit vulnerabilities. In 2023, a sophisticated attack campaign against multiple U.S. state agencies leveraged AI-powered tools to automate the process of identifying vulnerable systems and tailoring attacks based on the target's infrastructure. This incident showcases how AI can be used to scale and accelerate the initial stages of cyberattacks, potentially compromising many targets in a short period.

Case study: Star Health

The AI-powered ransomware attack on a prominent Indian healthcare provider in late 2024 serves as a stark example of the evolving cybersecurity landscape in the healthcare sector. This sophisticated attack leveraged AI across multiple stages of the CKC, demonstrating the potential for AI to enhance the effectiveness and stealth of cyber threats.

The attack began with an AI-driven *Reconnaissance* phase, where machine learning algorithms analyzed vast amounts of open source data to identify high-value targets within the organization. This allowed the attackers to quickly map out the healthcare provider's digital infrastructure and pinpoint vulnerabilities. The *Weaponization* stage utilized AI to create a custom ransomware variant specifically designed to evade the provider's security measures, showcasing the adaptability of AI-powered malware.

During the *Delivery* phase, the attackers employed highly convincing AI-generated phishing emails that mimicked internal communications, increasing the likelihood of successful infiltration. Once inside the network, the AI-driven malware demonstrated unprecedented autonomy, exploiting vulnerabilities and moving laterally while adapting its behavior to avoid detection by traditional security solutions. This adaptive capability made the attack particularly challenging to mitigate, as the malware could continuously evolve its tactics in response to defensive measures.

The impact of the attack was severe and multifaceted. Operationally, the healthcare provider experienced significant disruptions, with critical systems being encrypted and rendered inaccessible. This led to delays in patient care, potential compromises in data security, and substantial financial losses. The incident not only highlighted the vulnerabilities in the healthcare sector's cybersecurity infrastructure but also sparked industry-wide discussions on improving cyber defenses. The attack served as a wake-up call for other healthcare providers to review and enhance their cybersecurity postures, emphasizing the need for advanced AI-driven defense mechanisms to counter these sophisticated threats.

Challenges and considerations

While AI can significantly fortify an organization's defenses throughout the CKC, it is not without limitations and complexities. The following are the primary challenges and considerations that must be addressed to deploy AI responsibly and effectively:

- **Adversarial AI**: As cybercriminals increasingly leverage AI to create more sophisticated threats, organizations must continuously evolve their defensive AI capabilities to stay ahead in this technological arms race.

- **False positives**: Overzealous AI models can flag benign activities as threats, potentially leading to alert fatigue among security teams. Continuous refinement and human oversight are crucial to maintaining an effective balance.

- **Implementation complexity**: Deploying and integrating AI into existing cybersecurity frameworks requires significant investment, expertise, and organizational alignment. Organizations must carefully plan and execute their AI implementation strategies.

- **Ethical concerns**: The use of AI in cybersecurity raises important questions about privacy, data ownership, and the potential misuse of advanced surveillance capabilities. Organizations must navigate these ethical considerations carefully to maintain trust and compliance. AI's integration into cybersecurity, particularly within the CKC framework, offers powerful capabilities to anticipate, identify, and neutralize cyber threats. As demonstrated by real-world case studies, AI can significantly enhance an organization's security posture across all phases of an attack. However, the challenges associated with AI implementation underscore the need for a thoughtful, balanced approach to leveraging this technology in cybersecurity strategies.

An important but often overlooked aspect of integrating AI into cybersecurity is that AI can inadvertently amplify existing vulnerabilities. For instance, if sensitive information such as passwords, API keys, or private data is mistakenly exposed in plaintext and indexed by AI-driven systems, it significantly increases the risk of unauthorized disclosure. Users interacting with the AI prompt or query interfaces might inadvertently gain access to sensitive details, creating an expanded attack surface. Organizations must ensure robust data handling protocols, access controls, and comprehensive audits when implementing AI tools to prevent such inadvertent disclosures and maintain strict control over data visibility.

Having explored how AI reshapes attacker and defender capabilities in the CKC, we now turn to another transformative technology: **blockchain**. Although some CKC phases overlap conceptually with AI-driven protections, blockchain's unique structure and verification processes set it apart as a powerful complementary tool.

The role of blockchain technologies

Blockchain technology, celebrated for its decentralized, transparent, and tamper-resistant attributes, is emerging as a cornerstone in cybersecurity defenses. Its application across the CKC holds immense promise, enabling organizations to counter sophisticated threats with a combination of enhanced data integrity, threat visibility, and secure communication. By embedding blockchain into the CKC framework, security teams can reinforce every stage of the attack mitigation process.

Reconnaissance and weaponization phases

In the *Reconnaissance* phase, adversaries conduct intelligence gathering to identify potential weaknesses. Blockchain's decentralized architecture disrupts this process by distributing sensitive information across multiple nodes, thereby minimizing the risk of a single point of failure. Furthermore, blockchain offers tamper-proof logging, allowing organizations to record and monitor network scans or probing activities. These immutable records provide a forensic trail, enabling early detection of repeated reconnaissance attempts.

During *Weaponization*, where malicious payloads are developed, blockchain enhances security by supporting robust DevSecOps practices. Smart contracts ensure that only authorized software updates and code changes are deployed, reducing the risk of malicious actors embedding malware into legitimate applications. This automated integrity verification within the development pipeline bolsters defenses against malware weaponization.

Delivery and exploitation phases

As attackers transition to delivering malware, often via phishing or compromised URLs, blockchain can validate trusted sources. By maintaining a ledger of authenticated websites and email senders, blockchain-integrated systems can flag deviations and block malicious content before it reaches the intended target. This proactive approach prevents phishing emails from slipping through defenses and ensures secure communications.

In the *Exploitation* phase, attackers leverage vulnerabilities to gain access to systems. Blockchain-powered **decentralized identity** (**DID**) solutions provide an added layer of protection by validating user or system credentials against an immutable ledger. Unauthorized access attempts are promptly flagged, ensuring that only verified identities can interact with critical resources.

Installation phase

The *Installation* phase marks the attempt to establish persistence through malware or backdoors. Blockchain technology helps secure system integrity by maintaining an authoritative record of authorized software and configurations. Any deviation from this baseline, such as unauthorized installations, is immediately logged, flagged, and investigated. This tamper-proof mechanism ensures rapid identification and mitigation of installation attempts.

C2 phase

During the C2 phase, attackers aim to maintain communication with compromised systems. Blockchain disrupts this strategy by decentralizing the logging of network traffic and verifying communication channels. Anomalous patterns are detected in real time, preventing unauthorized connections to malicious servers. Smart contracts further enhance defenses by automating responses to suspicious activity, such as isolating compromised devices or revoking network access.

Actions on Objectives phase

In the final stage, where attackers execute their ultimate goals—whether data exfiltration or system sabotage—blockchain offers robust safeguards. Its immutable ledger ensures that data access and movement are meticulously tracked. Any unauthorized attempt to transfer or modify data is flagged, triggering immediate defensive measures such as encryption or isolation. This level of granularity enhances the organization's ability to detect and respond to malicious activities swiftly.

Types of blockchain-based cyberattacks

While blockchain strengthens cybersecurity in many ways, attackers have also evolved strategies to exploit vulnerabilities inherent in blockchain systems. Here are some notable types of blockchain-based cyberattacks:

- **51% attacks**: Malicious actors gain control of more than 50% of a blockchain's mining power or computing resources, enabling them to alter transaction records, double-spend cryptocurrency, or disrupt network operations.

- **Smart contract exploits**: Vulnerabilities in poorly written smart contracts can be manipulated to siphon funds or execute unauthorized actions.

- **Phishing in blockchain ecosystems**: Cybercriminals mimic wallet services or exchanges to steal private keys or credentials, often combining traditional phishing with blockchain-specific tactics.

- **Sybil attacks**: Attackers flood the network with fake nodes or identities, disrupting operations and manipulating consensus mechanisms.

- **Blockchain interoperability exploits**: Attackers exploit vulnerabilities in cross-chain communication protocols, compromising multiple interconnected blockchain systems.

- **Routing attacks**: By intercepting blockchain transactions, attackers can delay or manipulate data propagation, undermining the reliability of the network.

Real-world case study: the DAO exploit

The **Decentralized Autonomous Organization** (**DAO**) exploit in 2016 serves as a cautionary tale of the vulnerabilities inherent in blockchain ecosystems. The DAO, an Ethereum-based investment fund, used smart contracts to automate operations and democratize decision-making among its stakeholders. Despite its revolutionary potential, a flaw in the DAO's smart contract code left it susceptible to exploitation.

Attackers exploited the "recursive call vulnerability" in the smart contract, which allowed them to repeatedly withdraw funds before the system could update its ledger. Over $50 million worth of Ether was siphoned off within weeks, leading to widespread panic and a loss of confidence in the Ethereum network.

To mitigate the damage, the Ethereum community implemented a controversial "hard fork," creating a new version of the blockchain to return the stolen funds. This incident highlighted the critical importance of rigorous code auditing, robust smart contract development practices, and the need for a comprehensive incident response plan within blockchain ecosystems.

Challenges and considerations

While blockchain technologies offer significant potential to enhance cybersecurity and strengthen the CKC, their implementation is not without challenges. Organizations must navigate a range of technical, operational, and ethical considerations to fully realize the benefits of this transformative technology:

- **Scalability and performance limitations**: Blockchain systems often struggle with scalability, particularly in high-transaction environments, hindering real-time cybersecurity operations.

- **Integration complexity**: Incorporating blockchain into existing frameworks requires re-engineering systems, training personnel, and ensuring compatibility with legacy infrastructure.

- **Energy consumption**: Certain blockchain models, such as **Proof of Work** (**PoW**), are resource-intensive, posing ethical and environmental concerns.

- **Vulnerability of smart contracts**: Poorly written smart contracts are susceptible to exploitation, necessitating rigorous testing and validation.

- **Lack of standardization**: The absence of universally accepted blockchain standards complicates interoperability and governance.

- **Data privacy concerns**: Blockchain's transparency may conflict with data privacy regulations, such as GDPR, requiring a balance between immutability and compliance.

- **Cost implications**: Blockchain solutions, especially private ones, can be expensive to implement and maintain, posing challenges for smaller organizations.

- **Potential for misuse**: Just as blockchain strengthens defenses, adversaries can exploit it for malicious purposes, such as hosting malware or enabling anonymous transactions.

- **Legal and regulatory challenges**: Organizations must navigate the evolving regulatory landscape governing blockchain use, data protection, and financial transactions.

- **False sense of security**: Over-reliance on blockchain capabilities can lead to complacency, underscoring the need for a layered security approach.

Blockchain technologies bring transformative capabilities to the CKC, fortifying each phase with their inherent attributes of decentralization, transparency, and immutability. However, these systems are not impervious. Organizations must adopt a dual approach—leveraging blockchain for its security strengths while proactively addressing its vulnerabilities. As the cyber threat landscape evolves, integrating blockchain represents a vital step toward building resilient defenses that secure digital ecosystems against advanced adversaries.

While blockchain revolutionizes data integrity and trust in digital transactions, deception technologies take a different but complementary path, weaving illusions within cyberspace to mislead attackers, uncover their strategies, and strengthen cyber defenses.

Harnessing deception technologies for cyber defense

Deception technologies are an advanced cybersecurity strategy that involves deploying traps, decoys, and fake assets to mislead attackers, gather intelligence on their methods, and neutralize threats. By integrating deception technologies into the CKC, organizations can shift the advantage away from attackers, forcing them into controlled environments where they can be monitored and studied without compromising real assets. Deception technologies not only act as bait for malicious actors but also serve as a critical intelligence source for understanding attack patterns and improving future defenses. By leveraging these tools, organizations can maintain an active edge in the rapidly evolving threat landscape.

Deception technologies provide a proactive layer of defense, making the attacker's task more complex while allowing security teams to gain valuable insights into the **tactics, techniques, and procedures (TTPs)** of threat actors. This section explores how deception technologies can be used to disrupt attackers at each phase of the CKC.

Reconnaissance and weaponization phases

In the *Reconnaissance* phase, attackers scan the network, searching for vulnerable systems or services to exploit. Deception technologies can be employed by creating honeypots and honeynets—fake systems designed to look like real, vulnerable assets. These decoys lure attackers into interacting with them, thereby diverting their efforts away from real network resources.

Honeypots mimicking critical assets such as databases or servers can be strategically placed within the network to attract attackers. As attackers engage with these fake systems, security teams gain insights into their scanning techniques, tools, and IP addresses, allowing them to strengthen defenses on real assets. Any reconnaissance attempts on these decoy systems can trigger early warnings, giving security teams time to preemptively block or prepare for attacks.

In the *Weaponization* phase, where attackers are crafting malware or developing exploit code based on their reconnaissance, decoys can also serve as test environments for attackers. Deception technology allows the defender to control the interaction while capturing valuable information on the malware or techniques the attacker plans to use.

Deception-based environments can simulate vulnerable systems and log the attacker's efforts to weaponize their malware. This intelligence can then be shared with defense mechanisms across the network, enabling them to recognize and block the specific malware before it's used in an actual attack.

Delivery and Exploitation phases

In the *Delivery* phase, deception technologies continue to mislead attackers by providing decoy services or endpoints for them to interact with. For example, fake email addresses, servers, or endpoints can be deployed to catch phishing attempts or delivery of malware. Attackers may deliver malicious payloads to these decoys, thinking they have compromised real systems, while in reality, they interact with controlled environments designed to study their tactics.

Security teams can deploy deceptive mailboxes or web application forms that appear to be part of the real system. When attackers target these decoys with phishing emails or exploit kits, their actions are logged, and no actual system is compromised. This allows the organization to collect threat intelligence while preventing any real damage from occurring.

During the *Exploitation* phase, when attackers attempt to exploit a vulnerability to gain access, deception technology can play a critical role by creating fake vulnerabilities. Attackers are led to believe they've found a weakness in the system, but in reality, they are exploiting a fabricated vulnerability designed to trap and track their movements.

A deception system can emulate specific vulnerabilities that are common exploitation targets, such as unpatched software versions or misconfigured systems. Once an attacker attempts to exploit these, the security team can monitor the exploit techniques in real time, learning about the attacker's methods and patching similar vulnerabilities in actual systems before the real threat materializes.

Deception in the installation phase

In the *Installation* phase, where attackers attempt to install malware or backdoors into a compromised system, deception technologies can be invaluable. Fake software and systems can be deployed to trick attackers into installing their malware on decoy systems. These decoy systems then allow the organization to study the malware's behavior, gather forensic data, and develop mitigation strategies without risk to the real network.

Decoy endpoints can be configured to look like critical servers or user workstations, enticing attackers to install their malicious software. Once installed on the decoy, the malware's actions are tracked, and its payload is analyzed. This intelligence can then be used to update defenses and prevent the same malware from being installed on genuine endpoints.

Deception in the C2 phase

The C2 phase is critical for attackers to maintain control over compromised systems. Deception technologies can severely disrupt this phase by creating fake C2 channels that attackers believe are under their control. In reality, these channels are decoys, and every communication attempt made by the attacker is monitored by the defense team.

Honeypot-based C2 systems can mimic the behavior of a compromised system trying to communicate with a C2 server. When attackers attempt to issue commands through this decoy, their activities are recorded and analyzed. By simulating communication between malware and a C2 server, security teams can gain valuable insight into the attacker's infrastructure and techniques.

Furthermore, fake C2 channels can be used to feed false information back to the attacker, causing confusion and potentially delaying further stages of the attack. This disruption can buy valuable time for security teams to patch vulnerabilities or isolate compromised systems.

Actions on Objectives phase

In the *Actions on Objectives* phase, attackers attempt to achieve their final goals, whether that's exfiltrating data, damaging systems, or manipulating sensitive information. Deception technology can intervene here by using decoy files and databases to mislead attackers.

Deceptive filesystems can be configured to contain fake sensitive information. If an attacker attempts to exfiltrate these decoy files, security teams can track the data's movement and issue automated responses to block further actions. By leading attackers to decoy systems and files, organizations can prevent them from achieving their objectives and gain critical insights into their end goals.

In some cases, decoy systems can be set up to simulate a successful attack, only for the attacker to realize too late that they've stolen or manipulated non-sensitive, worthless data. This not only wastes the attacker's resources but also provides time for defenders to mount a real response.

Defending against cyberattacks with deception technologies

Deception technologies serve as a proactive defense mechanism, creating false realities to lure attackers and study their methods without compromising real systems. By employing decoy assets and environments, organizations can uncover hidden threats and mitigate risks effectively. Each deception technology targets a specific aspect of an attack and counters it. Let's take a look at how different deception technologies work for different types of cyberattacks:

Type of cyberattack	How deception technologies defend
Phishing attacks	Fake email accounts, decoy credentials, and honey users can lure attackers into exposing their tactics
Malware infections	Deploying sandbox environments and decoy systems traps malware, preventing it from reaching real assets
Ransomware attacks	Fake filesystems and decoy databases distract ransomware, analyzing its behavior while protecting real data
Insider threats	Honeytokens and decoy systems monitor and flag malicious actions by compromised or malicious insiders
Distributed Denial-of-Service (DDoS) attacks	Traffic redirection to decoy servers helps absorb and study attack patterns without disrupting actual services
Credential theft	Decoy credentials and fake login portals capture and monitor attempts to misuse stolen information
Lateral movement	False endpoints and network traps detect and contain attackers attempting to move laterally within a network
Zero-day exploits	High-interaction honeypots expose zero-day vulnerabilities and allow defenders to develop mitigation strategies
C2	Decoy C2 servers intercept and analyze attacker communications, breaking their operational chain
Supply chain attacks	Fake vendor accounts and supply chain components reveal threats targeting third-party dependencies

Table 9.1 – How deception technologies defend against different cyberattacks

Deception technologies not only act as bait for malicious actors but also serve as a critical intelligence source for understanding attack patterns and improving future defenses. By leveraging these tools, organizations can maintain an active edge in the rapidly evolving threat landscape.

Real-world case study of deception technology

In the lead-up to the 2017 French presidential election, Emmanuel Macron's campaign faced significant cyber threats, including phishing attacks and the dissemination of false information. To counter these challenges, the campaign implemented a proactive cybersecurity strategy that included elements of deception. This approach involved closely monitoring hacking attempts and deliberately feeding the attackers a mix of genuine and falsified documents.

This tactic aimed to sow confusion among the attackers and any entities attempting to exploit the stolen data. By interspersing authentic documents with fake ones, the campaign sought to undermine the credibility of any leaked information, making it difficult for adversaries to distinguish between what was real and what was fabricated. This method of cyber deception served as a defensive mechanism to protect the campaign's integrity.

Despite these efforts, on May 5, 2017, just two days before the election, a large cache of emails and documents from Macron's campaign was leaked online. The timing of the leak coincided with a mandated media blackout period, limiting the campaign's ability to respond publicly. However, due to the preemptive measures taken—including the seeding of falsified information—the impact of the leak was mitigated, as the authenticity of the documents was called into question.

The strategic use of deception technologies by Macron's campaign highlights the effectiveness of such measures in cybersecurity. By proactively introducing uncertainty and doubt into the attackers' efforts, the campaign was able to protect sensitive information and maintain its credibility during a critical period. This case underscores the potential of deception as a tool for defending against cyber threats in high-stakes environments.

Edge-driven cyber defense: integrating edge computing into the CKC

As cyber threats grow more sophisticated and pervasive, organizations must adapt their defenses beyond traditional network perimeters. One of the most promising developments in this area is the integration of edge computing into cybersecurity strategies. By moving computation and analysis closer to the data source—be it a sensor, industrial controller, or endpoint device—edge computing enables faster detection, quicker decision-making, and more agile responses. In the context of the CKC, this localized processing power can disrupt adversaries at every stage, from their earliest reconnaissance efforts through to their end goals. When thoughtfully implemented, edge computing allows defenders to identify anomalies, block malware, sever unauthorized communications, and thwart data exfiltration attempts before they cause serious damage.

This section explores how edge computing intersects with the CKC, examining how new levels of distributed intelligence, machine learning, and data processing at the network's fringes can bolster cybersecurity. We will also review various types of edge computing-based attacks, provide a real-world case study, and outline considerations and challenges that security teams must address as they adopt this emergent technology.

Integrating edge computing across the CKC

Edge computing introduces a strategic advantage by distributing detection and response capabilities directly onto endpoints, gateways, or field devices. Instead of shipping all data to a central location for analysis—which introduces latency—security tools embedded at the edge can rapidly detect and neutralize threats at their source.

Reconnaissance and Weaponization

In the initial stages, attackers survey their targets, scanning for vulnerabilities or weaknesses to exploit. Edge computing can disrupt this phase by placing anomaly detection tools directly on edge devices. For example, a router configured with local machine learning algorithms can spot unusual scanning attempts as they happen. Without relying on the cloud or a central server, these intelligent edge devices immediately flag anomalies—such as repetitive queries from suspicious IP addresses—enabling faster intervention.

During *Weaponization*, adversaries craft malware or exploit code tailored to identified vulnerabilities. By monitoring device behavior at the edge, defenders can detect early indicators of abnormal activity, such as unauthorized file downloads or unexpected port activity. This localized scrutiny can alert central security operations teams before the attackers finalize their payloads, making it easier to patch or reconfigure devices preemptively.

Delivery and Exploitation

Threat actors typically deliver their weaponized code via phishing emails, compromised websites, or Trojanized updates. With edge computing, gateways and local firewalls can run rapid threat intelligence scans on incoming payloads before they ever reach core systems. Suspicious links, files, or executables are flagged and blocked at the network's periphery, significantly narrowing the attack surface.

Should an adversary attempt to exploit a vulnerability, edge-based intrusion prevention systems can detect anomalous actions—such as unusual command executions or attempts to escalate privileges—in real time. By immediately raising alarms and blocking malicious requests at the device level, defenders prevent exploits from gaining traction in the environment.

Installation and C2

Once inside, attackers seek to establish persistence through backdoors or malicious implants. Edge computing counters this by continuously validating device integrity. If a smart sensor or connected camera suddenly tries to install unfamiliar software components, edge-based security processes can halt the operation, quarantine the device, or roll back unauthorized changes. These instant actions hinder attackers' attempts to embed long-term footholds in the network.

During the C2 phase, attackers try to maintain covert communication channels. Edge devices can inspect outbound traffic patterns to block suspicious requests to known malicious domains. By processing indicators of compromise locally, defenders can quickly sever C2 links, preventing attackers from orchestrating further moves inside the environment.

Actions on Objectives

Ultimately, attackers aim to steal data, disrupt operations, or damage critical systems. Edge-based security mechanisms can monitor data transfers, detect anomalous database queries, and identify unauthorized attempts to modify sensitive information in real time. For example, if a large file transfer suddenly appears bound for an unknown external host, local controls can immediately intercept and halt that action. This rapid intervention limits the window of opportunity for attackers to complete their mission.

Types of edge computing-based cyberattacks

While edge computing enhances defensive capabilities, it also opens the door to new attack vectors. Threat actors may target the very devices and infrastructure that constitute the edge layer, exploiting their distributed nature and reliance on emerging technologies.

Attack type	Description	Potential impact
Compromised edge devices	Hackers gain unauthorized access to IoT sensors, routers, or gateways to hijack local processing	Persistent footholds, altered data collection, and disrupted local operations
Malicious firmware updates	Adversaries distribute fraudulent firmware or patches to edge devices, embedding malicious code	Device instability, introduction of backdoors, and long-term compromise
Side-channel exploits	Attackers exploit resource constraints or CPU/GPU usage patterns on edge hardware to leak information	Data exfiltration, cryptographic key theft, and breach of device integrity
DDoS against edge infrastructure	Adversaries flood edge nodes with traffic to degrade localized processing and security operations	Service disruptions, degraded response times, and inhibited local detection

Table 9.2 – Edge Computing-based cyberattacks

The following real-world case study of the ArcaneDoor campaign illustrates how state-sponsored actors exploited weaknesses in perimeter edge devices, underscoring the critical importance of edge-driven cyber defense in detecting and responding to advanced persistent threats.

Real-world case study: ArcaneDoor campaign

The ArcaneDoor campaign, uncovered in April 2024, represents a sophisticated cyber espionage operation targeting government networks worldwide, primarily focusing on perimeter network devices such as Cisco **Adaptive Security Appliance (ASA)** devices. This state-sponsored attack, attributed to a threat actor known as UAT4356 or Storm-1849, demonstrates a high level of technical expertise and strategic planning in compromising critical infrastructure.

The campaign's initial access vector remains unknown, but the attackers leveraged two zero-day vulnerabilities in Cisco ASA devices: CVE-2024-20353 (a denial-of-service flaw) and CVE-2024-20359 (a persistent local code execution bug). These vulnerabilities allowed the threat actors to bypass authentication and gain a foothold in the target networks. Following the initial compromise, the attackers deployed two custom malware implants: Line Runner and Line Dancer. Line Runner utilized CVE-2024-20359 to load malware during system boot, while Line Dancer, an in-memory implant, leveraged CVE-2024-20353 to process C2 instructions.

Once established within the compromised networks, the ArcaneDoor operators demonstrated their ability to manipulate network traffic, modify device configurations, and maintain persistent access. This level of control allowed the attackers to conduct extensive espionage activities, including monitoring network communications, rerouting traffic, and potentially facilitating lateral movement within the targeted organizations. The campaign's focus on perimeter devices provided the attackers with a strategic vantage point for long-term intelligence gathering and data exfiltration.

The sophistication of the ArcaneDoor campaign is evident in its use of bespoke tooling and deep knowledge of the targeted devices, hallmarks of a well-resourced, state-sponsored actor. Subsequent analysis by security researchers has uncovered potential links to China, based on the infrastructure used in the attacks. Four of the five online hosts presenting the SSL certificate associated with the campaign's infrastructure were linked to Tencent and ChinaNet, prominent Chinese networks. Additionally, one of the attacker-controlled IP addresses was associated with anti-censorship software likely intended to circumvent China's Great Firewall.

The ArcaneDoor campaign underscores the critical importance of securing network perimeters and implementing a multi-layered defense strategy. Organizations are advised to regularly update and patch their network devices, implement rigorous monitoring systems, and conduct frequent security assessments to identify and mitigate potential vulnerabilities. The incident also highlights the ongoing trend of state-sponsored actors targeting edge devices and critical infrastructure, emphasizing the need for heightened vigilance and robust cybersecurity measures in both government and private sector networks.

Considerations and challenges

While edge computing provides powerful new defensive capabilities, it also introduces unique challenges:

- **Device diversity and complexity**: Edge environments often consist of heterogeneous devices, each with different capabilities and security requirements. Ensuring consistent security policies and patch management across this wide range of endpoints can be daunting.

- **Resource constraints**: Edge devices have limited processing power, memory, and battery life. Implementing advanced security analytics locally must be done efficiently to prevent degrading normal operations or shortening device lifespans.

- **Scalability and management overhead**: Managing thousands—or even millions—of distributed edge nodes can create a complex orchestration challenge. Security teams must have robust tools to deploy updates, configure policies, and monitor these systems at scale.

- **Privacy and data protection**: Storing and analyzing data at the edge can raise privacy concerns. Organizations must ensure that sensitive information processed locally is safeguarded, encrypted, and compliant with relevant regulations.

Edge computing brings vital speed, agility, and local intelligence to cybersecurity operations—a critical advantage when defending against advanced threats that evolve by the day. By embedding security directly at the data source, organizations gain a chance to detect, analyze, and respond to suspicious activity long before it escalates. Yet, as with any new technology, careful planning, robust architecture, and diligent oversight are required. Balancing performance, scalability, and privacy will remain an ongoing challenge. Done correctly, integrating edge computing into the CKC can significantly enhance an organization's posture, helping defenders keep pace with a complex and ever-changing threat landscape.

Synthetic data and PETs

While AI, blockchain, and edge computing can dramatically improve threat detection and response, they also intersect with another set of powerful tools: **synthetic data** and **PETs**. By generating artificial datasets and applying advanced privacy-preserving methods, organizations can train AI models and analyze data without exposing real sensitive information. However, these same capabilities can be misused by attackers to conduct sophisticated, low-visibility operations. In this section, we explore how synthetic data and PETs are leveraged in both offensive and defensive contexts across the CKC.

Types of synthetic data and PETs in cyberattacks

Synthetic data refers to artificially generated data that mimics real-world datasets while maintaining privacy and confidentiality. Its use in cybersecurity has grown due to the demand for robust data protection practices and the need for training machine learning models without exposing sensitive information. However, adversaries have started leveraging synthetic data generation techniques for malicious purposes. Types of synthetic data include the following:

- **Structured synthetic data**: Generated datasets that simulate structured formats such as financial transactions or database entries. These are often used in fraud detection systems but can also be exploited to train adversarial AI for bypassing fraud prevention algorithms.

- **Unstructured synthetic data**: Includes text, audio, or video data generated using techniques such as **generative adversarial networks (GANs)**. For instance, deepfake technology leverages unstructured synthetic data to create highly convincing but falsified media for social engineering or disinformation campaigns.

- **Time-series synthetic data**: Simulated data that reflects time-dependent patterns, such as network traffic or IoT sensor logs. Attackers use this to develop malware capable of mimicking normal activity patterns, making detection much more challenging.

PETs in cyberattacks

PETs are designed to enhance data privacy and security. While primarily intended for ethical use, attackers have adapted PETs for their operations, complicating detection and mitigation efforts. Key PETs include the following:

- **Homomorphic encryption**: Allows computations on encrypted data without decryption. Malicious actors use this to conceal their operations from forensic analysis, embedding attack payloads within encrypted streams.

- **Differential privacy**: Protects individual data points in datasets while allowing for aggregate analysis. Cybercriminals use differential privacy techniques to mask C2 communication patterns.

- **Federated learning**: Decentralized AI training on local data sources. Attackers leverage federated learning to orchestrate attacks on distributed systems without exposing the malicious training data to centralized detection mechanisms.

Real-world case study: Deepfake-assisted phishing campaign

In 2023, a multinational financial institution fell victim to a sophisticated cyberattack that leveraged deepfake technology and PETs to execute a fraudulent wire transfer of $35 million. The attackers employed GAN technology to create a convincing deepfake audio of a senior executive, which they used to manipulate an internal employee into approving the transfer.

The perpetrators demonstrated advanced technical skills by training their deepfake models on synthetic datasets. These datasets were meticulously crafted using publicly available audio recordings of the executive's speeches and interviews. This approach allowed the attackers to create a highly convincing impersonation that could bypass traditional security measures and exploit the human element of trust within the organization.

To further complicate detection and investigation efforts, the attackers utilized homomorphic encryption to conceal their operational details. This privacy-enhancing technique allowed them to process and manipulate encrypted data without decrypting it, significantly delaying the discovery of the breach and hindering attribution efforts.

The incident highlights the growing threat of deepfake technology in the financial sector, particularly when combined with other advanced cybersecurity techniques. Financial institutions are increasingly at risk from these sophisticated attacks that can bypass traditional security measures by exploiting human trust and leveraging cutting-edge AI technologies.

This case study underscores the urgent need for financial institutions to adapt their security protocols to address the emerging threats posed by deepfake technology and privacy-enhancing techniques. It emphasizes the importance of implementing multi-factor authentication systems, enhancing employee training on deepfake detection, and developing advanced AI-driven security measures to counter these evolving cyber threats.

Challenges

As powerful as synthetic data and PETs are for enabling privacy-respecting AI development and analysis, they introduce distinct challenges when misused by adversaries. The following list outlines the primary risks and limitations organizations must manage when integrating or defending against these technologies:

- **Detecting synthetic data usage**: Identifying adversarial uses of synthetic data is inherently challenging due to its similarity to legitimate data. Advanced AI tools are required to distinguish malicious patterns.

- **Encrypted traffic analysis**: PETs such as homomorphic encryption hinder traditional packet inspection tools. This creates blind spots in cybersecurity defenses.

- **Evolving techniques**: Attackers continuously refine their synthetic data generation and PET application methods, staying one step ahead of traditional defensive measures.

- **Resource constraints**: Advanced detection tools and AI-powered solutions to address synthetic data and PET misuse demand significant computational and financial resources, often unavailable to smaller organizations.

- **Collaboration barriers**: Sharing intelligence on PETs and synthetic data threats between industries is often hindered by privacy regulations, competitive concerns, and lack of standardization.

Post-quantum cryptography

As quantum computing advances, traditional cryptographic protocols face a looming existential threat. PQC offers algorithms resistant to the immense computational power of future quantum machines. In the context of the CKC, PQC helps protect sensitive data and communications from quantum-enabled adversaries who may eventually break today's encryption.

Types of PQC

The advent of quantum computing poses a significant threat to traditional cryptographic systems. PQC seeks to develop algorithms that remain secure against quantum attacks. Here are the primary types of PQC:

- **Lattice-based cryptography**: Lattice-based cryptography derives its security from problems related to lattices in multidimensional spaces. Algorithms such as **Learning With Errors (LWE)** and Ring-LWE are considered secure against both classical and quantum computers. These algorithms are efficient and versatile, making them a leading candidate for standardization.

- **Code-based cryptography**: Based on the hardness of decoding a general linear code, code-based cryptography has been around since the 1970s. McEliece encryption and Niederreiter schemes are prominent examples. While offering high security, these methods often require large key sizes, posing practical implementation challenges.

- **Hash-based cryptography**: This approach uses hash functions to create secure digital signatures. Lamport signatures and Merkle signature schemes are key examples. Hash-based cryptography is particularly suited for applications requiring a limited number of signatures, such as firmware signing.

- **Multivariate polynomial cryptography**: Multivariate polynomial cryptography relies on solving systems of multivariate quadratic equations, a problem that is computationally hard. Although offering the potential for PQC, these systems are not as mature as lattice- or code-based cryptography.

- **Isogeny-based cryptography**: This emerging approach is based on the difficulty of computing isogenies between elliptic curves. It has gained attention due to its small key sizes, making it a promising candidate for future cryptographic protocols.

Current progress in regulation and technology

Governments and standards bodies worldwide have recognized the urgency of addressing the quantum threat. Notable efforts include the following:

- **NIST Post-Quantum Cryptography Standardization project**: The **National Institute of Standards and Technology** (**NIST**) is leading efforts to evaluate and standardize PQC algorithms. As of now, lattice-based algorithms such as CRYSTALS-Kyber and CRYSTALS-Dilithium have emerged as front-runners.

- **EU Quantum Flagship**: The European Union's initiative focuses on developing quantum technologies, including cryptographic solutions, to secure critical infrastructure.

- **Industry-specific guidelines**: Sectors such as finance and healthcare are developing quantum-safe protocols to safeguard sensitive data against future quantum threats. For instance, Google has experimented with hybrid TLS using **X25519 + Kyber90s** to secure browser traffic.

Technological advancements

The private sector has made significant strides in implementing quantum-resistant systems:

- **Hybrid cryptographic systems**: Organizations are adopting hybrid models that combine classical and quantum-resistant algorithms to ensure a smooth transition.

- **Integration in IoT and edge devices**: Companies are exploring lightweight PQC algorithms suitable for resource-constrained environments such as IoT and edge devices.

- **Quantum-resistant cloud services**: Cloud providers are incorporating PQC to protect data at rest and in transit.

Real-world case study: Migration to PQC in financial institutions

JPMorgan Chase has successfully implemented a high-speed **quantum-secured crypto-agile network (Q-CAN)**, connecting two data centers over deployed fibers. This network utilizes **quantum key distribution (QKD)** to secure multiple independent, high-speed **virtual private networks (VPNs)** traversing a single 100 Gbps fiber.

The Q-CAN is part of JPMorgan Chase's dual remediation strategy, incorporating both PQC and QKD to prepare for the quantum era. This approach demonstrates the bank's commitment to maintaining trust and security in the global financial system.

The implementation of Q-CAN has shown that QKD integration can achieve enhanced security without sacrificing performance. The network's success in a production-level environment for financial services marks a significant advancement in quantum-safe communication.

JPMorgan Chase's initiative extends beyond the Q-CAN, as the bank has also established a third quantum node for testing next-generation quantum technologies applicable to banking and finance. This research platform will help the bank explore novel security features and prepare for future quantum threats.

The bank's proactive stance serves as a model for other financial institutions, highlighting the importance of early preparation for the quantum era. By investing in quantum-safe solutions now, JPMorgan Chase is not only protecting its current data but also securing its future operations in an increasingly complex cybersecurity landscape.

Challenges

PQC algorithms often come with increased computational and memory requirements. Organizations must balance security with performance, particularly in latency-sensitive applications such as financial transactions.

- **Interoperability**: Ensuring compatibility between quantum-resistant systems and legacy infrastructure is a significant challenge. Hybrid systems can ease this transition but add complexity to the implementation.

- **Key management**: The large key sizes required by some PQC algorithms pose challenges for storage and transmission, especially in resource-constrained environments.

- **Uncertainty in standardization**: While NIST's efforts provide a roadmap, the lack of finalized standards creates uncertainty for organizations planning long-term investments.

- **Awareness and skill gaps**: Adopting PQC requires specialized knowledge that many organizations currently lack. Building expertise in quantum-resistant technologies is crucial to successful implementation.

- **Cost implication**: Transitioning to PQC involves significant financial investment in hardware, software, and training. Organizations must justify these costs against the backdrop of an evolving quantum threat timeline.

Another key challenge in adopting PQC is managing its substantial resource requirements. PQC algorithms typically demand significantly higher computational power, increased memory, and storage capabilities compared to classical cryptographic methods. This can result in greater latency, particularly challenging for performance-sensitive applications such as real-time financial transactions and IoT devices with limited resources. Organizations must strategically plan their infrastructure upgrades, balancing enhanced security with potential performance trade-offs. Additionally, the financial investment required for hardware enhancements, software integrations, and staff training can be considerable, especially for smaller entities or resource-constrained environments. Careful cost-benefit analysis and phased deployments can help mitigate these impacts, ensuring organizations adopt PQC solutions effectively without overwhelming their existing operational capacities.

PQC represents a critical advancement in the fight against emerging cybersecurity threats. While promising, its adoption requires careful planning, significant investment, and a willingness to navigate the challenges of implementation. As this chapter demonstrates, emerging technologies such as PQC will play an integral role in fortifying the CKC and securing the future of digital systems.

Summary

This chapter explored how advanced technologies are being integrated into the CKC to address increasingly sophisticated cybersecurity challenges. It examined the roles of AI, blockchain, deception technologies, edge computing, and PQC, highlighting their impact across the kill chain's stages. AI is pivotal in automating threat detection, anticipating malicious actions, and neutralizing attacks in real time. Blockchain, with its decentralized and tamper-resistant nature, strengthens security by ensuring data integrity and facilitating secure communication. Deception technologies play a crucial role in misleading attackers with decoys and traps, allowing defenders to gather critical intelligence while protecting real assets. Edge computing enhances threat detection and response by processing data closer to its source, and PQC provides resilience against emerging quantum computing threats.

This chapter also featured real-world examples, such as ransomware incidents powered by AI and blockchain-based security implementations, demonstrating the practical applications of these technologies. It addressed key challenges, including adversarial AI, complex implementation processes, and ethical considerations, while offering strategies to overcome these obstacles. By adopting these advanced tools, organizations can enhance their ability to detect, defend, and adapt to modern cyber threats across all phases of the CKC. Concluding with an emphasis on the transformative potential of these technologies, the chapter set the stage for the next discussion on legal and ethical aspects of the CKC, which will focus on the responsible deployment of these innovations in alignment with regulatory, ethical, and privacy considerations.

Further reading

Following are the case studies and articles we discussed in this chapter as well as some bonus reading materials for you to check out:

- *AI-Powered Ransomware Attack on a Healthcare Provider*: Explores a 2024 ransomware attack on an Indian healthcare provider, highlighting the use of AI-driven malware to disrupt operations and compromise sensitive data (`https://www.cyberpeace.org/resources/blogs/research-report-ai-powered-ransomware-attack-on-a-healthcare-provider`)

- *2017 Macron Email Leaks*: Details the leak of over 20,000 emails from Emmanuel Macron's campaign just before the 2017 French presidential election, examining allegations of Russian involvement and the leak's impact (`https://en.wikipedia.org/wiki/2017_Macron_e-mail_leaks`)

- *Defending Against ArcaneDoor: How Eclypsium Protects Network Devices*: Analyzes the ArcaneDoor cyber espionage campaign targeting Cisco network devices, providing insights into malware implants and strategies for improving network security (`https://eclypsium.com/blog/defending-against-arcanedoor-how-eclypsium-protects-network-devices/`)

- *State-Sponsored Espionage Campaign Exploits Cisco Vulnerabilities*: Covers the ArcaneDoor campaign that exploited vulnerabilities in Cisco firewalls for espionage purposes, urging organizations to patch devices to mitigate risks (`https://www.infosecurity-magazine.com/news/stateespionage-campaign-cisco/`)

- *Understanding the ArcaneDoor Campaign*: Explains how the ArcaneDoor campaign targeted perimeter network devices using sophisticated tactics, emphasizing defense-in-depth strategies to mitigate risks (`https://cymulate.com/blog/understanding-the-arcanedoor-campaign/`)

- *China-Linked Hackers Suspected in ArcaneDoor Cyberattacks*: Investigates links between Chinese state-sponsored actors and the ArcaneDoor campaign based on infrastructure analysis (`https://thehackernews.com/2024/05/china-linked-hackers-suspected-in.html`)

- *Tackling State-Sponsored Cyber Espionage in Network Perimeters*: Explores technical details of the ArcaneDoor campaign, including zero-day vulnerabilities exploited by malware implants and recommendations for securing perimeter devices (`https://blog.qualys.com/vulnerabilities-threat-research/2024/04/24/arcanedoor-unlocked-tackling-state-sponsored-cyber-espionage-in-network-perimeters`)

- *ArcaneDoor: State-Sponsored Malware Targeting Cisco Devices*: Provides an overview of the ArcaneDoor campaign, focusing on its use of zero-day vulnerabilities to compromise Cisco ASA devices for espionage activities (`https://www.conquer-your-risk.com/2024/04/26/arcanedoor-state-sponsored-malware-targeting-cisco-devices/`)

- *Researchers Discover Links Between China and ArcaneDoor Campaign*: Attributes the ArcaneDoor campaign to Chinese actors based on infrastructure analysis, linking SSL certificates and anti-censorship tools to Chinese networks (`https://fieldeffect.com/blog/researchers-discover-links-china-arcanedoor`)

- *Treasury Warns Banks About Rising Deepfake Fraud Threats*: U.S. Department of Treasury warns financial institutions about increasing fraud schemes involving deepfake technology, including manipulated identity documents and AI-generated customer profiles (`https://www.americanbanker.com/news/treasury-warns-banks-of-rising-deepfake-fraud-threat`)

- *Network Traffic Anomaly Detection with Machine Learning*: Explores the use of machine learning techniques for detecting anomalies in network traffic, enhancing cybersecurity measures (`https://eyer.ai/blog/network-traffic-anomaly-detection-with-machine-learning/`)

- *AI in Cybersecurity: Key Case Studies and Breakthroughs*: Highlights significant case studies and breakthroughs in the application of AI to cybersecurity, showcasing its potential in threat detection and prevention (`https://eastgate-software.com/ai-in-cybersecurity-key-case-studies-and-breakthroughs/`)

- *AI in Threat Detection*: Discusses the role of AI in enhancing threat detection capabilities, improving response times, and automating security processes (`https://www.paloaltonetworks.com/cyberpedia/ai-in-threat-detection`)

- *AI Threat Intelligence*: Examines how AI is revolutionizing threat intelligence by processing vast amounts of data to identify and predict potential security threats (`https://bigid.com/blog/ai-threat-intelligence/`)

- *Role of Artificial Intelligence in Threat Detection*: Explores the various ways AI is being integrated into cybersecurity systems to enhance threat detection and response mechanisms (`https://www.sangfor.com/blog/cybersecurity/role-of-artificial-intelligence-ai-in-threat-detection`)

- *AI in Cybersecurity: Snorkel AI Perspective*: Provides insights into how Snorkel AI's technology is being applied to improve cybersecurity measures through advanced AI techniques (`https://snorkel.ai/ai-in-cybersecurity/`)

- *AI in Cybersecurity: TechMagic Blog*: Offers a comprehensive overview of AI applications in cybersecurity, including threat detection, risk assessment, and automated response systems (`https://www.techmagic.co/blog/ai-in-cybersecurity/`)

- *AI Arms Race: Artificial Intelligence in Military Technology*: Discusses the implications of AI in military technology and the potential for an AI arms race between nations (`https://foreignpolicy.com/2023/04/11/ai-arms-race-artificial-intelligence-chatgpt-military-technology/`)

- *Artificial Intelligence in Healthcare: Opportunities and Challenges*: Explores the potential applications and challenges of AI in healthcare, including diagnostic tools, treatment planning, and patient care (`https://www.ncbi.nlm.nih.gov/pmc/articles/PMC10030838/`)

- *Weaponization of AI: The New Frontier in Cybersecurity*: Examines how AI is being weaponized in the cybersecurity landscape, presenting both new threats and defensive capabilities (`https://www.hkcert.org/blog/weaponisation-of-ai-the-new-frontier-in-cybersecurity`)

- Cisco AI Security Study (2023): *AI in Cybersecurity: Key Case Studies and Breakthroughs* (`https://eastgate-software.com/ai-in-cybersecurity-key-case-studies-and-breakthroughs/`)

- Deloitte Blockchain Report (2024): *Blockchain to Combat Data Tampering* (`https://www2.deloitte.com/content/dam/Deloitte/us/Documents/Advisory/us-advisory-digital-assets-banking-and-capital-markets-regulatory-digest-october-2024.pdf`)

Get This Book's PDF Version and Exclusive Extras

UNLOCK NOW

Scan the QR code (or go to `packtpub.com/unlock`). Search for this book by name, confirm the edition, and then follow the steps on the page.

Note: Keep your invoice handly. Purchase made directly from Packt don't require one.

Legal and Ethical Aspects of the Cyber Security Kill Chain

Cybersecurity professionals know that technical prowess alone isn't enough – we must also navigate a maze of legal requirements and ethical expectations. The **Cyber Kill Chain (CKC)** framework, originally developed by Lockheed Martin to outline the stages of a cyberattack, provides a useful lens to examine where these legal and ethical issues arise.

This chapter expands on each phase of the CKC – **Reconnaissance, Weaponization, Delivery, Exploitation, Installation, Command and Control**, and **Actions on Objectives** – highlighting the key legal implications and ethical considerations at every step.

We'll also dive into global cybersecurity laws (such as GDPR, CCPA/CPRA, HIPAA, PIPL, LGPD, etc.), discuss hot topics such as "hacking back," data privacy, duty of care, and proportionality, and learn from recent case studies (SolarWinds, MOVEit, and Colonial Pipeline). Throughout, the tone is practical and hands-on, aimed at helping you apply these insights in real-world security operations. By the end, you should have not only a clearer understanding of the laws and ethics governing cybersecurity but also concrete recommendations and checklists to guide your actions.

Next, we will learn about each phase of the CKC and the legal and ethical considerations for them.

Phase 1: Reconnaissance – legal and ethical considerations

Reconnaissance is the attacker's information-gathering stage. The intruder selects a target and scouts for weaknesses – researching the company, employees, technologies in use, and potential vulnerabilities. Common tactics include scanning network ranges, scraping LinkedIn or social media for employee information, reading job postings (to infer software in use), and even phishing for initial information. Defenders also perform reconnaissance, in a sense, by gathering threat intelligence on potential attackers or running scans on their own systems (or those of partners, with permission) to find exposures.

Legal implications

Simply put, *unauthorized reconnaissance can cross legal lines.* Port scanning and footprinting a network without permission exist in a gray area – in many jurisdictions, mere scanning isn't explicitly illegal, but it can be a step toward illegal access. In the U.S., the **Computer Fraud and Abuse Act (CFAA)** makes it unlawful to access a system without authorization; while a basic scan may not trigger it, aggressive scanning or using credentials improperly certainly can. European laws (such as the UK's **Computer Misuse Act 1990**) more directly criminalize even preparatory acts – for example, attempting to identify vulnerabilities in someone else's system could be seen as an *"unauthorized access attempt."* Gathering publicly available data (**Open source Intelligence (OSINT)**) is generally legal, but even here there are limits: web scraping might violate a website's terms of service or data protection laws if personal data is involved. Under **data privacy regulations** such as **General Data Protection Regulation (GDPR)**, if you, as a defender, collect personal information during threat intelligence (say, compiling a list of names and emails of suspected threat actors or scraping hacker forums), you may need a lawful basis to process that data. It sounds odd, but even criminals have personal data that could be protected by law – a European company collecting information on an attacker in, say, Russia might technically be processing personal data. The key is to focus on information that is necessary and avoid gratuitous personal details to stay compliant with privacy laws. For defenders conducting **penetration testing** or **red teaming** exercises, it is crucial to obtain explicit authorization and clearly define the engagement scope. Ethically, red teams should simulate realistic adversarial actions, including scenarios where employees might unintentionally expose sensitive organizational information, such as posting proprietary code or security keys publicly. Such simulations help organizations prepare more robust defensive strategies. Operating without proper documented permission, including clearly articulated ethical boundaries and stakeholder approvals, could lead to legal repercussions despite good intentions. Therefore, always ensure comprehensive written authorization and stakeholder agreement before conducting any adversarial emulation activities.

A famous example is the case of penetration testers who were arrested in Iowa in 2019 for testing courthouse security outside of agreed hours – miscommunications led to legal consequences despite their good intentions. The lesson: always ensure your reconnaissance and testing activities are authorized in writing.

Ethical issues

Reconnaissance raises an important question: *how far should you pry?* For attackers, this phase is inherently unethical – it's preparation for a crime, often involving deception (such as pretexting to get information) or privacy invasion. As defenders, however, we also have ethical boundaries. When conducting employee security awareness tests, for instance, it might be tempting to dig into employees' social media to craft ultra-realistic phishing emails. But is it ethical to use someone's personal posts or information about their family to trick them? Many would argue that security testing should avoid *overly personal social engineering*, as it can feel like a betrayal of trust or an invasion of privacy. In another scenario, let's say you discover an external researcher or hacker poking around your network (reconnaissance from the outside). Ethically, you should respond proportionally – maybe by monitoring or blocking the IP – but not by aggressively hacking them back (we'll discuss "hacking back" soon). During reconnaissance, a defender might set up **honeypots** or fake data to attract attackers. That's generally considered ethical and legal (on your own network) as a defensive measure, but you must ensure any monitoring of attacker behavior doesn't accidentally expose the real private data of innocents. The guiding ethical principle in recon is **respect for privacy and proportionality**. Gather information needed to secure your assets, but don't collect extraneous personal data *"just because."* Also, be transparent within your organization – for example, if you monitor employee network activity for threats, employees should be made aware of this in a fair policy. It's about finding the balance between vigilance and respect for individual rights.

Practitioner tips for the Recon phase

Before you implement reconnaissance techniques or respond to incoming scans, consider these practical guidelines to ensure legal compliance and ethical oversight during this early stage of the CKC:

- **Always get authorization for scans/tests**: If you're assessing someone else's systems (e.g., a vendor or partner), obtain written permission. Even when conducting internal tests or aggressive scanning activities, it's essential to define and document the scope clearly. Explicit permissions must be obtained from appropriate stakeholders within the organization. This includes identifying the specific teams and systems involved and ensuring all parties agree to the extent of testing. Clear documentation of authorization helps avoid potential misunderstandings or legal complications and reinforces the ethical conduct of internal security assessments.

- **Use threat intel ethically**: Collect threat data that helps defend (IoCs, threat actor TTPs, etc.), but avoid storing unnecessary personal information. For any personal data you collect (names, emails of threat actors, etc.), protect it and have a legitimate need.

- **Employee awareness testing**: When simulating phishing or social engineering, do it to teach, not to humiliate. Don't leverage sensitive personal details about employees in your faux attacks – keep it professional.

- **Monitor within legal limits**: If you deploy honeypots or monitoring tools, ensure they're on networks you own. If you somehow gain intel that an external system is targeting you, involve legal/authorities rather than snooping on that external system yourself.

- **Document your activities**: Keep logs of your own recon (such as pentest reports and scan results) in case questions arise. This helps demonstrate that you acted within the allowed scope and for legitimate defense purposes.

Having uncovered how attackers scout their targets and how defenders legally safeguard their intelligence efforts, let's see what happens when reconnaissance data transforms into an actual cyber weapon.

Phase 2: Weaponization – legal and ethical considerations

In the *Weaponization* phase, the attacker takes the intelligence gathered and crafts or selects a payload to exploit the target. In classic terms, this might be pairing a malware **remote access trojan** (**RAT**) with an exploit for a known vulnerability, and wrapping it into a deliverable file (such as a malicious Office document or PDF). Essentially, it's the stage of building the "weapon" – malware development, exploit writing, and packaging tools – before sending it off. Defenders typically aren't "weaponizing" anything, but they might be preparing countermeasures or exploit detection tools at this stage if they anticipate an attack (e.g., creating YARA rules for known malware).

Legal implications

Creating malware or exploit code exists in a tricky legal space. Additionally, cybersecurity practitioners should be aware of export control laws such as the **International Traffic in Arms Regulations** (**ITAR**), which govern the export of certain exploit tools and cybersecurity software classified as *munitions*. Practitioners must verify that sharing, transferring, or exporting these tools internationally complies with applicable laws and regulations to avoid legal risks. Writing code, even if capable of malicious use, is generally protected as an expression or for legitimate research. However, *distributing* malware or exploit tools can cross legal lines. Many countries outlaw the supply of "hacking tools" to unauthorized parties. For instance, the UK's **Computer Misuse Act** has Section 3A, which makes it illegal to make, adapt, supply, or offer any article intending to be used to commit or facilitate a computer offense. That could include selling a password cracker or a malware toolkit. The intent matters – tools shared among security researchers for defense are one thing, but selling a botnet kit on a forum will get you in trouble. In the U.S., there isn't a direct "malware-making" law, but creators of malware can be charged as conspirators or under statutes for damage caused by their tools. A real example is the author of the Mirai botnet malware (used to knock large swaths of the internet offline in 2016) who was prosecuted in the U.S. not for writing code per se, but for the damage and criminal conspiracy that code enabled. Also, export control laws can apply. Advanced encryption or exploit software might be seen as munitions; shipping them overseas without a license might violate regulations (the old "crypto as a munition" issue). If an enterprise security team develops specialized offensive tools for red teaming, they must be careful about how those are stored and used – if such a tool was leaked and misused, the organization could face liability if it was negligent in safeguarding it. For defenders, weaponization might involve creating custom scripts to disable an attacker's malware or to simulate attacks for testing. That's fine if kept in-house. But be wary of any action that could be seen as building a weapon to use outside your network. For instance, some well-meaning defenders have considered crafting their own retaliatory malware to infect the attacker back – that definitely would violate laws (unauthorized access, likely).

Ethical issues

The ethics of weaponization primarily concern researchers and defenders in how they handle exploit knowledge. For example, if you discover a zero-day vulnerability in software, you essentially have a "weapon." Ethically, responsible disclosure is expected: you inform the vendor privately, give them a chance to patch, and *do not* use it for personal gain or hold it secretly hoping to exploit competitors. There have been contentious debates about releasing **proof-of-concept (PoC)** exploit code publicly. On one hand, releasing a PoC can help educate and pressure vendors to fix issues; on the other hand, it arms less-skilled attackers who might use it immediately. A practitioner should weigh the greater good – many follow the norm of disclosing exploit details only after a patch is available, and even then, sometimes only conceptually rather than a fully working attack script. Within a company, if the security team develops attack simulations, they should ensure these tools are used strictly for defense testing and are well controlled. Ethically, there's also the concept of **do no harm** – even if you have the capability to strike back at an attacker or to create malware as a demonstration, consider the potential unintended consequences. One illustrative incident was the release of the *Adobe Photoshop ransomware* as a so-called ethical experiment by a researcher – it encrypted files but was meant to teach a lesson. Many argued that it was unethical, as it intentionally caused damage (even if reversible) to unwitting users' machines. The bottom line is that developing exploits and tools should be aimed at strengthening defenses, not punishing attackers or showing off prowess at the expense of others. From an attacker's perspective, weaponization is obviously unethical – designing a tool to harm or steal is malicious intent. But even state actors justify their cyber weapons as "necessary for national security," raising ethical questions on the global stage about the proliferation of cyber weapons (for example, the EternalBlue exploit developed by the NSA leaked and caused the WannaCry and NotPetya outbreaks, causing billions in damage – was stockpiling that exploit ethical given the eventual harm when it escaped?). As a cybersecurity practitioner, you may not control nation-state decisions, but you can control your own approach: prioritize responsible research and disclosure over secretly stockpiling exploits.

Practitioner tips for the Weaponization phase

With an effective payload, attackers are ready to move from planning to action. The next challenge is ensuring their weapon actually reaches the intended target – ushering us into the *Delivery* phase:

- **Handle exploits with care**: If you discover a vulnerability, first check whether the affected organization offers an official bug disclosure or bug bounty program. Using these sanctioned channels ensures responsible disclosure. Always involve your legal and management teams in planning and executing vulnerability disclosures to mitigate risks and adhere to ethical standards and best practices.

- **Follow responsible disclosure guidelines**: Notify vendors, CERTs, and so on. Don't hoard zero-days for "maybe one day" offensive use; it can backfire legally and ethically.

- **Control your tools**: Keep any custom attack tools or malware samples in a secure environment. Treat them like live ammo. Limit access to those who truly need it for defense testing.

- **No DIY retaliation malware**: However tempting, designing a "reverse virus" to hit back at your intruder is a no-go. It is not only illegal, but ethically, you could hit an innocent by mistake. Leave offensive operations to law enforcement or the military with proper authority.

- **Training and simulations**: It's okay to weaponize *simulated attacks* in controlled exercises. Just ensure all participants know it's an exercise (or if it's a surprise drill, that at least leadership and legal sanctioned it). Also, avoid simulations that might accidentally spill out (for example, a fake malware that could propagate beyond the test group). Contain your training "weapons."

- **Ethical red teaming**: If you're on a red team, remember the goal – improve the organization's security. Don't get carried away with clever attack techniques that might cause real damage (such as wiping data or causing production outages) just to "win" the engagement. Follow the agreed rules of engagement strictly.

With an effective payload now in hand, the attacker's next step is to send it into the target environment—bringing us to the critical phase of delivery.

Phase 3: Delivery – legal and ethical considerations

In the *Delivery* phase, the attacker delivers the weaponized payload to the target environment. This could be through phishing emails with malicious attachments, a drive-by download on a compromised website, a USB stick dropped in a parking lot, or even physical delivery (in the case of an insider or someone dropping hardware). Essentially, it's the transmission of the exploit/malware to the victim. For defenders, this is where mail filters, web proxies, and other perimeter defenses come into play to intercept malicious deliveries. It's also where security awareness is key (to stop that employee from plugging in the random USB or clicking that rogue link).

Legal implications

Once we reach delivery, the attacker is usually clearly committing a crime. Sending a malicious payload to a system without permission is an unauthorized action. Under laws such as the CFAA in the U.S., even *attempting* to access a computer without authorization is illegal – so firing off a spear-phishing email with a malware attachment to an employee is definitely an attempt at unauthorized access (exploitation will be the actual access, but the delivery itself is part of the criminal act). Besides anti-hacking statutes, other laws might apply; for example, if the delivery mechanism is an email with fraudulent elements, it could constitute **wire fraud** or fall under anti-phishing and cyber-fraud laws. In some jurisdictions, if malware is delivered across national borders, it can trigger international law issues – an attacker in country A sending malware to systems in country B is violating country B's cybercrime laws and could possibly be subject to extradition (though that's often tricky in practice). For defenders, legal concerns in this stage revolve around how you *intercept* and *inspect* content. Consider an organization that inspects incoming emails and web traffic for threats – this often involves scanning attachments or even opening them in a sandbox, and doing deep packet inspection on network traffic. Privacy laws, such as data protection regulations, require that you have a legitimate reason to monitor communications. Luckily, "protecting the network from malware" is usually a very defensible legitimate interest or accepted practice, but organizations should still disclose to users (in corporate policies) that their communications may be scanned for security. In some countries, there are specific laws about intercepting communications. For instance, in the U.S., the **Electronic Communications Privacy Act (ECPA)** allows service providers to intercept and filter communications if it's necessary to protect the provider's rights or property – which covers anti-malware scanning by an employer on their own network. Just ensure your employees or users are informed via an acceptable use policy that says, for example, *"All network traffic may be monitored for security."* Looking from another legal angle, what if a defender sets up a *sinkhole* or *fake server* to attract malicious deliveries? An example could be creating a dummy email account that attackers are likely to target, so you can capture their phishing attempt. That's legal on your own systems and often done (think of it like a spam trap). Just don't extend this to tricking the attacker into delivering something to *their* detriment beyond your boundaries (again, hacking back concerns). Also, if an organization decides to test employees by delivering simulated malware (such as a benign file that looks real to test whether IT staff respond correctly), be sure there's internal agreement – sending even fake "malware" without coordinating could raise internal accountability issues if someone panics and calls law enforcement or if it disrupts work.

Ethical issues

The *Delivery* phase for attackers is where they actively involve unwilling participants, which is clearly wrong. Phishing often preys on human trust – attackers might send emails that look like they were sent from a colleague or a trusted service. This betrays trust and can cause harm beyond just the malware (for example, a phishing email that impersonates a CEO might cause panic or confusion within a company, even if the malware is caught, eroding morale). For defenders, ethics come into play in how we train and test our users on these deliveries. **Phishing simulations** are a common practice to teach employees to be wary. However, these must be done ethically. There have been cases where companies sent very cruel phishing tests – one company sent an email to employees offering a bonus or financial relief, which was fake and intended to see who would click. Employees felt this was a heartless trick, and it damaged trust in leadership. The ethical approach is to make simulations *realistic but fair*. Don't exploit employees' emotions unnecessarily (such as a phishing test that mimics a layoff notice or something equally sensitive). Another ethical aspect is respecting user autonomy: if an employee fails a phishing test, use it as a coaching opportunity, not a shame campaign. The goal is to educate, not punish or embarrass. Additionally, consider **supply chain ethics** – many deliveries come through third-party channels (for example, the SolarWinds case involved delivering a compromised software update to thousands of organizations). If you're a software provider, you have an ethical duty to ensure your update delivery mechanism is secure, because your software delivery could become the attack vector for your customers (as it did in SolarWinds). After that incident, many companies re-evaluated how they do secure code signing and update servers, recognizing an ethical responsibility to customers to not be a conduit for malware. Finally, if you catch an attempted delivery (say, your filters catch a malicious email), an ethical question is: should you inform the source if it seems like an innocent party was leveraged? For instance, an attacker might compromise someone's email and use it to send you malware. If you recognize that, ethically, you might want to alert that sender that their account is compromised. This kind of **good karma** action helps the broader community, though it's not legally required.

Practitioner tips for the Delivery phase

Once malicious software is prepared, attackers must deliver it to the victim's environment. This crucial step connects the planning stage to the real-world impact on targeted systems. Here are some tips to remember when responding to the *Delivery* phase:

- **Strengthen your human firewall**: Conduct phishing training and simulations regularly, but keep them ethical. Use scenarios relevant to your business but avoid cruel tricks. If someone clicks a test link, provide instant feedback and learning materials rather than public shaming.

- **Legal vetting of monitoring**: Work with your legal team to ensure your email and web monitoring practices are compliant with laws and communicated in policy. If you operate in regions such as Europe, make sure employees are aware (e.g., via consent or at least clear notice) that their incoming/outgoing communications will be scanned for threats.

- **Intercept wisely**: Use technical controls such as attachment sandboxing and URL filtering to catch malicious deliveries. Ensure that any quarantined content is handled carefully – if it's an employee's personal email in the mix, treat it with respect for privacy while still protecting the organization.

- **Supply chain due diligence**: If you rely on third-party software or services, inquire about how they secure updates and deliveries. On the flip side, if you provide software, invest in securing your build and update pipeline – consider it both a legal risk (liability if your negligence leads to customer breaches) and an ethical obligation.

- **Incident response preparation**: Have a plan for when a malicious delivery is detected. This includes technical steps (isolating affected machines) and communication steps. For example, if a partner organization inadvertently sent you malware (because they were compromised), decide whether you will reach out to warn them – it's often the right thing to do and can build goodwill. Document any such outreach or notifications.

After learning how malicious payloads enter a victim's system, we now turn to the moment those payloads take effect: *Exploitation*.

Phase 4: Exploitation – legal and ethical considerations

Exploitation is the moment the delivered payload takes action and actually exploits a vulnerability to gain unauthorized access or control of the target system. For example, the user opens that malicious attachment and it executes code that exploits a software vulnerability to give the attacker a foothold. Or a user visits a hacked website, and the site triggers an exploit in their browser. In the *Exploitation* phase, the attacker's code is running on the victim system, breaking the security (such as bypassing authentication, elevating privileges, etc.). From here, the attacker can install further backdoors (leading into the *Installation* phase) or start pivoting deeper. Defenders view exploitation as the "breach point" – it's where intrusion detection systems hopefully catch suspicious behavior, and where system hardening and patching (or lack thereof) come into play.

Legal implications

By the time exploitation occurs, there's no ambiguity: a crime has been committed. Unauthorized access to computer systems is illegal virtually everywhere. The exact statutes can vary – CFAA in the U.S., various national cybercrime laws elsewhere – but they all prohibit breaking into systems or exploiting security flaws without permission. If the exploitation causes damage (which could be as simple as crashing a system or as severe as deleting data), additional charges such as computer damage or sabotage could apply. For instance, if an exploit triggers malware that encrypts files (ransomware), that's destruction, or at least alteration of data – an offense on top of unauthorized access. An interesting legal point: what if the attacker is an *insider* using legitimate credentials but for illegitimate purposes? This phase could still be at play (an insider exploiting their access to go where they shouldn't). In some jurisdictions such as the U.S., after the Supreme Court's Van Buren decision in 2021, simply misusing access might not fall under "exceeding authorized access" depending on how the law is interpreted, but it can violate other laws (trade secret theft, fraud, etc.). Companies often have employees sign acceptable use agreements, so insider exploitation would at least be a breach of contract and likely other computer misuse laws that cover insiders. For defenders, legal aspects of exploitation involve *how you allow or block exploitation*. For example, deploying patches for known exploits is actually part of your legal "duty of care" – regulators might view not patching critical vulnerabilities as negligence (especially if it leads to a breach of personal data protected by laws such as GDPR or HIPAA). There have been cases where companies were fined for not preventing known exploits (e.g., Equifax's 2017 breach involved an unpatched Apache Struts vulnerability; while their fines were mostly for the breach impact, underpinning that was a failure to patch known exploitable flaws).

Here's another consideration: when you detect an exploit attempt, what are you legally allowed to do? You can certainly block the traffic or quarantine the system. You can collect forensic evidence from your own systems (logs and memory dumps) without issue – that's your data. But if you somehow have the capability to trace the exploit back to the attacker's system (say the malware is connecting back to an IP and you have an exploit for *their* server), you must not "counter-hack" without law enforcement involvement. Law enforcement agencies, with proper warrants, *can* sometimes exploit the attacker's infrastructure to shut it down (for example, the FBI may deploy network investigative techniques in criminal infrastructure given court authorization). But if a private company did the same on its own – say, exploited a vulnerability in the attacker's command server to take it offline – that company would itself be committing an unauthorized access crime. The concept of **active defense** versus **illegal hacking back** looms large here (we will explore it in a dedicated section later). The short takeaway is that after you've been exploited, you *cannot* justifiably exploit the attacker in return without running afoul of the law. Stick to defensive containment and evidence gathering on your turf.

Ethical issues

Exploitation marks the point where trust is broken – the attacker has violated the integrity of a system. Ethically, this is indefensible from the attacker's perspective; they are knowingly causing harm or violating privacy. For defenders, how we respond ethically is critical. One ethical aspect is **disclosure**. If exploitation results in a breach of data, there's an ethical (and often legal) duty to disclose that breach to those affected. We'll cover breach notification in the *Actions on Objectives* phase since it's closely tied to data exfiltration, but even at the moment of exploitation, the wheels of responsibility start turning. For example, if you discover that an attacker exploited a vulnerability to access a database of customer information, the clock likely starts on when you need to notify regulators or users (GDPR says within 72 hours to regulators, and many U.S. state laws say notify customers "without undue delay"). It's ethical to notify promptly and transparently – hiding an exploit or breach is considered unethical (as well as legally punishable in some cases).

The Uber 2016 breach is a notorious example: attackers exploited a weakness and stole data on 57 million riders and drivers. Uber's leadership at the time chose to pay the attackers to delete the data and labeled it a "bug bounty" rather than report the truth. This cover-up was unethical and ultimately led to legal action against Uber and a conviction of the CSO in charge for obstruction. So, when exploitation happens, *ethical leadership means owning up to it*. Another ethical issue is how you treat suspected attackers caught in the act. Suppose your system catches an exploit attempt and identifies a specific employee's machine as the source (maybe an insider threat or their credentials were used). Ethically, you should investigate fairly – confirm the facts before accusing. It could be malware using their account, not the person themselves. We owe individuals due process even internally. Also, consider **systemic ethics**: exploitation often takes advantage of the most vulnerable – such as outdated systems in a hospital (as the WannaCry ransomware did). Society generally agrees it's especially egregious to exploit targets that can risk lives (hospitals and emergency services). As a defender in such sectors, there's an ethical weight to your mission: you know attackers have no qualms about hitting you, so you have to work even harder to patch and protect because real lives could be at stake (not just data).

There's also an ethical dimension in the cybersecurity community: when we find a new exploit (zero-day), do we always immediately weaponize it? Reputable researchers abide by ethics and don't use it maliciously. But what about intelligence agencies or contractors who find exploits? They might justify using them for national security. This introduces the debate of "offensive ethics" – outside the scope of a corporate practitioner, but as citizens, it's a relevant discussion (e.g., should agencies disclose exploits to vendors to protect the public or keep them secret to use against enemies?). The ethical stance in the private sector should be clear: if you find or buy any exploit information, use it to protect and fix, not to further exploit others.

Practitioner tips for the Exploitation phase

After malicious content lands on a victim's system, the real damage begins once that code executes. Exploitation is the moment attackers gain unauthorized control, making it the true start of the breach. Here are some tips to remember when planning for the *Exploitation* phase:

- **Patch, patch, patch**: The best way to handle exploitation is to prevent it. Maintain a robust patch management program. Know your high-risk systems and prioritize critical updates. If a critical vulnerability is published (such as those in exchange servers or VPN appliances in recent years), assume attackers will exploit it within days (or hours) and act accordingly.

- **Intrusion detection**: Deploy host-based and network-based **intrusion detection systems (IDSs)** to catch exploit behavior – for example, unusual process launches, memory injections, or known exploit signatures. Ethically, ensure your detection methods don't violate user privacy more than necessary – for example, scanning the memory of servers is fine, but recording keystrokes of users "just to see" is not okay unless there's cause.

- **Evidence preservation**: When an exploit is detected, start preserving evidence immediately in a forensically sound manner. This is both for your incident investigation and in case law enforcement needs it. It's legal and ethical to collect data from your own systems to understand the breach. Just be mindful if that data includes the personal information of employees or customers – secure it and limit access to breach responders.

- **Containment with care**: If a user's machine is exploited, isolate it from the network, but also consider the user. Inform them in an appropriate way so they know what's happening (e.g., "*Your workstation appears to have been compromised; we're going to take it offline and give you a clean one.*"). This respects the employee and avoids panic or rumors.

- **Don't retaliate in kind**: No matter how angry you are at the attacker, resist any urge to "hack them back." Instead, gather what information you can (IP addresses, malware samples, etc.) and share that with authorities or industry threat intel networks. Contribute to the bigger defense community rather than taking offense into your own hands.

- **Plan for disclosure**: If the exploit succeeded in breaching data or systems, loop in your legal and communications team early. Start drafting breach notification statements if needed. Being proactive and honest is ethically right and often legally required. It's better coming from you than from a journalist or the attacker posting your data on X/Twitter.

Once the attacker gains initial access by exploiting vulnerabilities, the focus shifts to installing and embedding persistent malware within the compromised system.

CKC phase 5: Installation – legal and ethical considerations

In the *Installation* phase, the attacker solidifies their presence by installing malware on the compromised system – typically, a persistent backdoor or implant. After exploiting a vulnerability and gaining initial access, they want to ensure they don't lose that access. This could involve installing a remote access trojan, adding new user accounts, or modifying startup scripts so their malware runs on reboot. Essentially, the attacker is *establishing persistence*. For example, malware might drop copies of itself in multiple directories or schedule tasks to reconnect to the attacker's server. From a defender's perspective, the *Installation* phase is where you might detect unusual changes on systems (new services, odd files, registry changes, etc.) and where **endpoint detection and response (EDR)** tools can flag suspicious activity.

Legal implications

Unauthorized installation of software on someone else's system is certainly illegal – it's part and parcel of the unauthorized access offense. In fact, many laws treat the placement of malware as a distinct crime: for instance, planting a keystroke logger could be an offense of unauthorized surveillance on top of the hacking itself. If the malware installed is **ransomware**, then it additionally becomes extortion once the attacker demands payment to remove it. If it's spyware stealing data, it could violate wiretap laws or privacy laws in addition to computer crime laws. From a legal standpoint, one interesting aspect is when law enforcement themselves install something on a suspect's computer – a "lawful malware", if you will (often called **network investigative techniques** or **NITs**). They require a warrant and are heavily regulated because even for the good guys, installing software on someone's system without consent is a big deal legally. For everyday defenders, a question might arise: can we ever install something on an attacker's system as a defense? The answer is generally no (unless it's your system that the attacker is using – e.g., they compromised your server and you, in response, installed a beacon on that server to track them, which is still dicey). In the corporate context, sometimes defenders deploy beacon documents or files ("canary" files) that phone home when opened, to see whether data gets accessed by an attacker. Those aren't really installations on the attacker's systems; rather, they trick the attacker into installing or running something on their side. Legally, this is murky but, generally, if the attacker willingly executes it, some argue it's fair game. Still, such active defenses walk a fine line. In most cases, organizations avoid any action that could be construed as delivering code to an attacker. Instead, they focus on eradicating what the attacker installed on their own systems. It's fully legal for you to remove or neutralize malware on your machines – that's within your rights. One caveat: if the malware is logging information or there's a chance to recover stolen data from it, coordinate with incident responders before blindly deleting it. There have been cases where companies immediately wiped malware, losing forensic evidence that could have been used to catch the perpetrator or understand the breach scope. Legally, destroying evidence can hamper law enforcement, though companies won't be penalized for it if done as part of good-faith remediation (intent matters). Still, the best practice is to image and preserve before cleaning up.

Ethical issues

The *Installation* phase from the attacker's view is deeply unethical – they're effectively planting a hidden bug in someone's digital home. It's akin to a burglar secretly installing a camera in your house after breaking in. For defenders, ethics revolve around how much we invade privacy or disrupt systems to get rid of the malware. One ethical dilemma that sometimes comes up is: should we ever leave the malware running for a bit to study it (or maybe track the attacker's activity) versus immediately removing it?

If the malware is not causing active harm and you have law enforcement involved, you might consider letting it run under controlled observation to gather intel on the attacker (for instance, to see whether they'll reveal additional compromised systems or their intentions). But you have to weigh that against the risks – the attacker could do more damage while you watch. Ethically, our first duty is to protect the organization and its data subjects, not to play cyber detective. So generally, it's better to contain/remove malware as soon as feasible, after collecting the necessary evidence. Another ethical aspect: when cleaning an infection, be transparent internally about what happened. Some companies are tempted to quietly remove malware and not tell anyone, perhaps to avoid embarrassment. But an ethical security culture calls for informing relevant teams and management about the incident, maybe even the whole company if appropriate (especially if it stemmed from, say, an employee falling for phishing – you can frame it as a learning opportunity without blame). Openness helps reinforce vigilance across the organization.

You have to also consider the ethics of communication with customers/partners. If an attacker installed a backdoor and possibly had ongoing access for a while, the ethical move is to let affected parties know. For example, during that period, if they might have accessed customer data or used your systems to attack others, you should inform those customers or partners. This overlaps with breach notification duties. A case in point: in the *Sunburst* malware incident (the SolarWinds supply chain attack), once discovered, SolarWinds had to ethically and legally inform all customers who took the tainted update that their systems might have a backdoor installed. Even though SolarWinds was a victim, they had an obligation to help customers respond (which they did, alongside CISA guidance). On the defender side, installing software is usually done for good (such as EDR agents and monitoring tools), but even that needs an ethical check – ensure those tools themselves are secure and don't violate privacy more than necessary. We've seen some security tools that collected more information than they should or introduced vulnerabilities of their own. Keeping security agents updated and minimizing their data collection to just security metadata is an ethical approach to balance security and user privacy.

Practitioner tips for the Installation phase

With initial access achieved, attackers aim to stay in control. The *Installation* phase places the malicious tools firmly within the environment—ensuring the intruder can return even if the initial exploit is discovered. Here are some tips to remember when planning for the *Installation* phase:

- **Hunt for persistence**: Once an incident is confirmed, do a thorough sweep for any backdoors the attacker may have installed. This includes looking at autorun entries, new services, scheduled tasks, unfamiliar user accounts, SSH keys in servers, and so on. Many EDR tools can show persistence mechanisms. Removing these is critical so the attacker doesn't re-enter after you think you've cleaned up.

- **Collaboration with IT**: Removal of malware might require reimaging machines or restoring from backups. Work closely with IT to do this in a controlled way. Ethically, if an employee's device is involved, reassure them – often, people feel guilty or scared they did something wrong. Support them and explain the steps being taken.

- **Check legal reporting duties**: At this stage, if you haven't already, determine whether the compromise triggers any regulatory notifications. For instance, under **Health Insurance Portability and Accountability (HIPAA)** (health sector law in the U.S.), if malware was installed on a system containing patient data, it could be considered a breach unless you prove no data was accessed. Often, you won't know without deeper analysis, so err on the side of reporting if required by law. The legal counsel should be making these calls, but security provides the facts.

- **Improvement for next time**: Every malware installation that succeeds should be a lesson. Conduct a post-incident review: how did it evade our defenses? Did our antivirus miss it (and have we submitted the sample to our antivirus vendor now)? Did our users not report the odd behavior (maybe need more training to recognize signs of compromise)? Use this to strengthen your security controls.

- **Ethical use of "good" malware**: You might hear of concepts such as "white worms" – benign worms that remove bad ones. Historically, things such as the Welchia worm tried to remove the Blaster worm from systems. As a defender, don't deploy such things in your environment unless extremely controlled and sanctioned (and never on others' networks). Generally, it's not recommended due to unpredictable side effects. Instead, rely on standard cleanup tools from reputable sources.

- **Communication**: If the incident is significant, work on an internal communication (and external if needed) plan. Being honest that "*X malware was found and removed*" and providing guidance on what employees or customers should do (e.g., change passwords if a keylogger was present) is both ethical and bolsters trust. It shows you're on top of it and care about protecting stakeholders.

After establishing a foothold on the network, attackers look to maintain ongoing control—a task achieved through the **Command and Control (C2)** stage.

Phase 6: Command and Control – legal and ethical considerations

In the C2 phase, after installing a foothold, attackers need a way to control the compromised systems remotely – this is where C2 comes in. The malware (now running persistently on victim machines) will reach out to the attacker's servers for instructions or to exfiltrate data. Typical C2 methods include connecting to a hardcoded domain or IP address, using DNS queries to receive commands, opening an encrypted tunnel to a control server, or even using legitimate services (such as hiding commands in X/Twitter posts or Slack messages). The attacker can then issue commands to the infected machine, such as navigating the network, collecting files, or executing additional tools. For defenders, the C2 phase is often when you detect the breach, because unusual outbound connections or traffic patterns might trigger alerts. Techniques such as network traffic analysis, DNS monitoring, and anomaly detection are crucial here. It's also where threat intel (known malicious domains/IPs) and sinkholing operations come into play.

Legal implications

From the attacker's side, maintaining a C2 server is part of the ongoing unauthorized access crime. If the C2 servers are in another country, it complicates law enforcement but doesn't change the illegality. Often, multiple jurisdictions get involved (e.g., the attacker might be in country A, C2 servers in country B, and victims in country C). International cooperation is needed to take down C2 infrastructure. This is where treaties such as the **Budapest Convention on Cybercrime** facilitate law enforcement partnerships, enabling things such as cross-border server seizures or data sharing. For example, during the disruption of the Emotet botnet in 2021, authorities from numerous countries coordinated to seize servers and redirect C2 traffic, effectively issuing an update to remove the malware from victims' computers – all done lawfully via joint warrants in each jurisdiction. That kind of global takedown is legal only with proper authority in each place. If a private company tried to, say, use DDOS on the C2 or hack it, that would be illegal vigilantism. However, companies do assist sometimes by providing technical information or sinkhole capabilities to law enforcement. A **sinkhole** is a server that security researchers or authorities set up to mimic C2, luring infected machines to connect to it so their traffic can be observed or cut off from the real C2. Setting up a sinkhole for a domain that an attacker was using often requires legal steps – typically, a court order to get the DNS registrar to point that domain to the sinkhole server. Private security companies often work in collaboration with the government on this. Legally, defenders are fine to monitor their own outbound traffic to detect C2. However, if you want to monitor C2 communications beyond your network, you must hand off to law enforcement. Intercepting communications on someone else's server or network without permission would violate wiretapping laws. Also, a lot of C2 traffic is encrypted. In some cases, companies might be tempted to use their decryption capabilities (such as if they decrypt SSL at the proxy) to inspect C2 data. That's legal for your own traffic (with user consent via policy), but you couldn't demand an ISP give you someone else's traffic data – that requires warrants by investigators. Another legal consideration is during an incident response: you might isolate an infected system but still allow it to talk to C2 in a controlled way to see what commands are being issued (for intel). Doing so is legal on your side, but if you start responding through that channel (such as tricking the attacker via the malware), you're wading into possible "communication interception" or unauthorized access territory. Usually, responders will simply block C2 or allow it to talk to a sinkhole, rather than engage interactively with the threat actor via the malware. Let law enforcement handle any engagement (sometimes they do – pretending to be a bot to trace actors, etc., with proper authority).

Ethical issues

C2 is where attackers often use innocent third parties as part of their scheme. They might host C2 on compromised servers of legitimate businesses, essentially using someone else's property to facilitate crime. This is double victimization – those server owners often have no clue their systems are rallying points for a botnet. As defenders, an ethical step when we discover such cases is to notify those third parties. For example, if you identify that an attacker's malware in your network is reaching out to a specific company's server (which appears to be compromised and acting as a C2), it's good ethics (and good karma) to reach out through channels such as a CERT or directly if you know contacts there. Many in the security community do quietly inform others of breaches (*"Hey, I see your web server at IP X.Y.Z is beaconing to my compromised machine – you might want to check it."*). It helps break the kill chain beyond your own walls and builds trust among peers. From the defender's ethical standpoint, another consideration is how much you monitor egress traffic. To catch C2, deep inspection of outbound traffic is often needed (such as decrypting SSL using a proxy). Ethically, this intersects with user privacy – employees might be using the company internet for personal things (such as checking personal webmail, banking, etc.). If you decrypt and log everything, you could capture sensitive personal information. The ethical approach is to minimize the retention of content and focus on security metadata or known bad indicators. Some organizations choose not to decrypt traffic to sites that likely carry personal communications (e.g., banking sites) to respect privacy and only decrypt traffic to unknown or less personal destinations. It's a balance each organization must consider. Ethically, we also must consider the *proportionality of response* at the C2 stage. If we detect data being exfiltrated via C2, one might consider drastic measures such as shutting down the entire network segment or internet connection to stop it. That might be necessary in extreme cases, but it can also disrupt business or services to customers. The ethical choice is to contain the threat in a way that minimizes impact on innocent parties. For example, isolate just the affected servers rather than cutting off the whole company's internet (unless the breach is so widespread that you have no choice). Communication with the attacker is another ethical area – some companies have engaged ransomware attackers via C2 chat portals (basically negotiating). While paying a ransom is controversial ethically, simply talking is not unethical per se and can buy time or clarity. But one should never misrepresent oneself (e.g., pretend to be law enforcement in those chats) – keep it honest if you engage, or better yet, have law enforcement or professional negotiators do it.

Practitioner tips for the C2 phase

Now that an attacker has established a hidden foothold, they must maintain stealthy oversight. C2 is what transforms isolated malware into a fully remote operational platform. Here are some tips to remember when planning for the C2 phase:

- **Network egress filtering**: Implement strict outbound firewall rules and web filtering. Many organizations allow internal systems to initiate connections out freely – attackers exploit this. Lock down outbound traffic to only what's needed (e.g., servers shouldn't make random web requests if they normally don't; user PCs likely don't need to talk to every port out there). By restricting this, you can prevent malware from successfully reaching its C2 in the first place.

- **DNS monitoring and sinkholing**: A lot of malware uses DNS to locate C2 (e.g., by querying specific domain names). Use DNS firewall services or internal DNS logging to catch anomalies. If you see devices querying domains that are known bad or random-looking, investigate. You can even sinkhole (redirect) known malicious domains to a safe internal server to see which machines try to connect – this identifies infected hosts. Ethically, ensure this doesn't accidentally disrupt non-infected users (generally, sinkholing a clearly malicious domain is a low risk since only malware should be using it).

- **Threat intelligence**: Subscribe to threat intel feeds for C2 indicators (IP addresses, domain names, file hashes of tools, etc.). Integrate these into your **Security Information and Event Management (SIEM)** or firewall block lists. When using intel, consider the source – use reputable feeds to avoid false positives that could block legitimate traffic. If you do block something based on intel, have a process for users to report if they were prevented from accessing a legitimate site, so you can quickly adjust.

- **Incident coordination**: When an active C2 is detected, it's time to pull in your incident response team fully. This may involve legal, PR, and certainly management because decisions (such as whether to shut down systems or engage with attackers) might need leadership approval. Have predefined criteria for when to do a major containment, such as disconnecting from the internet. That way, you're not panicking – you're executing a plan.

- **Law enforcement involvement**: Strongly consider contacting law enforcement, especially if it's a significant breach or involves ransomware/extortion. They can offer guidance, and sharing C2 indicators with them might contribute to a larger takedown. It's both ethical (help catch the bad guys beyond just your incident) and practical (they might have additional intel for you). Remember to preserve those C2 logs; they could be evidence.

- **Communicate internally**: Let your IT and users know that you've blocked certain traffic or taken certain systems offline to combat C2. For example, if you block all outbound SSH because an attacker was using it, tell your developers why their SSH connections are failing suddenly. Transparency helps maintain trust during the chaotic response period.

Armed with a steady link into the victim's environment, the attacker can finally pursue the main goal—be it data theft, sabotage, or extortion—in the *Actions on Objectives* phase.

Phase 7: Actions on Objectives – legal and ethical considerations

Actions on Objectives is the endgame of the attack – the attacker, having established control, now works to achieve their ultimate goals. Depending on the attacker's motive, this phase can look very different. For a state-sponsored espionage attacker, the objective is likely data exfiltration – quietly stealing sensitive information (documents, databases, and emails) and transferring it out via the C2 channels. For a ransomware gang, the objective is data encryption and extortion – they may both steal data (double extortion scheme) and then encrypt your files, followed by a ransom demand. For a hacktivist, the goal might be data destruction or public release to cause embarrassment or make a political point. For an **Advanced Package Tool (APT)** attacking critical infrastructure, the objective could even be sabotage – such as altering industrial control systems to cause physical damage (think of Stuxnet causing centrifuges to tear themselves apart, or an attacker opening a dam's floodgates). Essentially, *Actions on Objectives* is whatever malicious outcome the attacker ultimately delivers upon the target. From the defender's perspective, this is where the damage manifests and where the incident response must mitigate the impact. It's also the phase that triggers most legal obligations such as breach notifications, and tough ethical decisions (e.g., whether to pay the ransom or not).

Legal implications

Different types of actions correspond to different laws. **Data theft (exfiltration)** triggers data breach laws and privacy regulations. For instance, under the EU's GDPR, if personal data is accessed or stolen, the organization must report the breach to regulators within 72 hours and potentially to affected individuals. Regulators can investigate and impose fines if the organization doesn't adequately protect the data. Similar requirements exist under U.S. state laws (such as California's breach notification law), Canada's **PIPEDA**, **Australia's Privacy Act**, India's upcoming laws, and so on. So, legally, once you know data was taken, the clock is ticking on your duties to notify. Moreover, stolen data could lead to lawsuits from customers or partners if they suffer harm. We've seen class-action lawsuits after breaches (for example, customers suing Target after its 2013 breach for failing to secure their card data). **Ransomware encryption** triggers a host of issues too. While paying a ransom is not outright illegal in most places, there are legal nuances: if the ransomware gang is under economic sanctions (say, believed to be in a sanctioned country or on a sanctions list), paying them could be illegal (violating **Office of Foreign Assets Control** (OFAC) regulations in the U.S., for example). In 2020, the U.S. Treasury explicitly warned that companies facilitating ransom payments might violate sanctions if the recipient is blacklisted. So, organizations have to consider that. Additionally, some sectors have to report ransomware incidents to authorities – for instance, the banking sector in some countries must report if they paid a ransom, and in the U.S., critical infrastructure owners will soon have to report cyber incidents and ransom payments under new federal rules. **Destructive attacks or sabotage** could engage laws beyond cyber – for example, if an attack on a power grid causes outages, there could be public safety laws, and it might even be considered an act of terrorism. Internationally, if one nation-state causes destructive effects on another's infrastructure, it could breach international law (violating the principle of non-intervention, possibly even amounting to the use of force under the UN Charter if the damage is severe). That's a bit beyond a practitioner's immediate concerns, but it shows gravity. On the defender side, when dealing with *Actions on Objectives*, legal considerations are about *response and aftermath*. If personal data is involved, coordinate with legal on breach disclosure. If it's intellectual property stolen, maybe involve law enforcement for theft (the FBI often helps companies on IP theft cases, especially if a foreign state is involved). Also, preserve evidence of what was taken or destroyed; it may be needed for legal proceedings or insurance claims. Speaking of insurance, many companies have cyber insurance that might cover certain losses or ransom payments – but insurers often require that you follow certain practices and notify them promptly. Failing to involve law enforcement or paying a ransom without insurer consent could affect coverage, so legal teams will check those policy obligations too. In summary, legally, this phase turns a technical incident into a reportable, accountable event where regulators, law enforcement, and possibly courts become involved.

Ethical issues

This phase is where ethical leadership and decision-making are most visibly tested. Let's talk first about ransomware and extortion. When faced with a ransom demand – pay to get decryption and/or to prevent leaked data – organizations are in a moral bind. Additionally, organizations must be aware that certain jurisdictions explicitly restrict ransom payments. It is critical to consult legal counsel to understand and comply with local regulations regarding ransom payments, as violating these laws could lead to severe legal consequences.

On the one hand, paying could quickly restore operations (important if public services or lives are at stake, such as a hospital ransomware case) and could protect customers' data from being leaked. On the other hand, paying funds criminal activity and encourages more attacks (and there's no guarantee the criminals won't leak or come back anyway). Ethically, most law enforcement bodies and governments urge not to pay, aligning with the principle of not negotiating with criminals. However, when you're in the hot seat, the ethical calculus can shift if, say, patients can't get care or an entire community's fuel supply is halted (as in the Colonial Pipeline case, they paid about $4.4 million to DarkSide to get back up quickly). Many ethicists would still argue that alternatives (such as good backups for restoration and emergency continuity plans) are the better path if at all possible, rather than funding crooks. This is why having those backups and plans is often pointed to as an ethical responsibility of companies – you owe it to society to not be so fragile that your only resort is enriching criminals. Another ethical facet is transparency with those affected. If customer data was stolen, how candid are you in your public statements? Ethically, full transparency (within the bounds of accurate knowledge) is advised. That means if you know data was taken, say so, and let people know what kind of data, so they can take protective actions (such as changing passwords or watching credit reports). Spinning or downplaying is unethical and can backfire legally too. Consider the early statements in some breaches that were cagey, only for it to emerge later that things were worse – that erodes trust badly. It's better to say *"We're still investigating the scope, but here's what we know, and here's what we recommend you do now."* than to issue a vague *"We had a cybersecurity incident, no evidence of misuse"* statement when, in fact, data is streaming out to dark web markets. Duty of care is a huge ethical theme here: did we do enough to prevent this outcome? In the aftermath, leaders should ethically assess this. Maybe it's not for public self-flagellation, but internally, take responsibility. If mistakes or neglect contributed (such as failing to act on an alert that preceded a big breach), acknowledge that within the team and fix the process. Ethically, companies should also consider the harm to those whose data was taken. Offering support such as credit monitoring to affected individuals, or providing clear assistance channels, is part of making it right.

Some companies also choose to openly share their learnings from attacks to help others – that's an ethical contribution to the community (e.g., Maersk's and Norsk Hydro's openness after major attacks has been praised; they lost a lot, but they shared how it happened and how they recovered, benefiting others). When attacks have a societal impact (such as Colonial Pipeline's fuel disruption), ethical leadership means cooperating fully with government efforts to mitigate public harm, even if it means revealing embarrassing details about the breach. Finally, consider employees: sometimes after a breach, there's a temptation to find someone to blame (the person who clicked, the admin who misconfigured, etc.). Ethically, avoid scapegoating. Unless there was gross misconduct, most breaches are systemic failures. Support your team; don't throw a sacrificial lamb. The stress employees go through in recovering from an incident is enormous – showing compassion and giving them the resources and rest they need is the right thing to do.

Practitioner tips for the Actions on Objectives phase

With a remote pipeline into the compromised network, attackers can begin realizing their true intentions. Whether it's stealing sensitive data or orchestrating further destructive actions, this is where motivations become tangible. Here are some tips to remember when planning for the *Actions on Objectives* phase:

- **Activate full incident response plan**: This stage is crisis mode. Follow your pre-written incident response plan, which should include technical steps (such as disconnecting affected systems and securing backups), communication steps, and stakeholder notifications. Having a checklist here is invaluable because it's easy to get overwhelmed.

- **Breach notification prep**: If personal data is impacted, draft the notification letters and regulatory reports quickly. Use plain language for customers – tell them what happened, what information of theirs was involved, and how you're helping (e.g., providing credit monitoring, a hotline for questions, etc.). Run these drafts by legal to ensure compliance with specific laws (each jurisdiction has slightly different required info to include).

- **Engage law enforcement and external experts**: By this point, if you haven't already, involve law enforcement. Also, consider hiring incident response consultants or forensic firms if the situation is beyond your in-house capacity. Many regulators look favorably if you show you brought in independent experts – it demonstrates diligence. These experts might uncover clues your team missed; they also help quantify what data was stolen for reporting purposes.

- **Decide on ransom pragmatically**: If hit by ransomware, convene an emergency decision group (typically, executives, legal, infosec, and possibly board members if the situation warrants) to assess whether to pay the ransom. Provide them with critical input: Do we have working backups? How quickly can we restore without paying? Is the stolen data of an especially sensitive nature (e.g., personal health information versus general office documents)? Additionally, gather intelligence on the threat group – are they known to provide decryptors upon payment? Are they on any sanction lists?

In some jurisdictions, paying a ransom to sanctioned entities or terrorist organizations is explicitly prohibited, and violating these regulations could result in severe legal consequences, including fines, sanctions, and potential criminal charges. For example, in the U.S., North Carolina and Florida have banned public entities from paying ransoms, and federal laws prohibit transactions with sanctioned entities as per OFAC guidelines. The UK is also considering a ban on ransom payments by public bodies, and the Counter Ransomware Initiative coalition advises against paying ransoms to cybercriminals.

Always consult legal counsel to understand the legal exposure and regulatory implications before making any payment. Usually, negotiators or law enforcement can assist in providing this crucial intel. This decision isn't purely technical; it encompasses business, legal, and ethical considerations. Document the rationale for whatever course of action you take, as you may need to justify it later to regulators, courts, or insurance providers.

- **Public relations** (PR): Work closely with PR or corporate communications to manage public messaging. In a crisis, rumors fly; having a clear message helps. Be truthful and empathetic. For example, acknowledge the inconvenience or fear this incident may cause customers, and express commitment to security improvements. Avoid speculative statements ("*We think it was China!*" unless confirmed, for instance). Stick to what you know and what you're doing about it.

- **After-action analysis**: Once the fire is out, do a retrospective. This should cover technical fixes (patch that hole, improve that logging, etc.), process improvements (earlier detection and better training), and also a review of how well the incident response went. Were roles clear? Did anyone hesitate to escalate? Did we follow the legal steps correctly? Use this to refine plans and maybe conduct a drill later to practice any new procedures. Share key learnings with the whole IT/security team and even the company if appropriate – a post-mortem report can increase awareness and support for security initiatives (often after an incident, the budget for security gets a boost, so be ready with a list of what you need).

- **Support your people**: Incidents are draining. Recognize the extra hours the team put in. Consider some time off or at least a less intense period to recuperate. Upper management should thank the responders – it goes a long way for morale. And if an employee made an error that led to the incident, turn it into a training point, not an HR firing (unless it was egregious negligence or malicious). Most of the time, the blame lies in the process or lack of safeguards, not an individual being malicious. Promote an atmosphere of learning and improvement rather than fear of punishment, so people report issues promptly in the future.

By now, we've seen how each stage of a cyberattack unfolds. It's time to place these phases within the broader context of international laws and regulations—rules that often dictate how organizations must respond when breaches occur.

Global legal frameworks in cybersecurity and data privacy

Modern cybersecurity work doesn't occur in a vacuum; it's regulated and guided by a patchwork of laws around the world. Understanding the major legal frameworks is crucial for practitioners, especially as attacks and data routinely cross international borders. Let's break down some key global and regional laws and regulations that relate to the CKC phases, particularly where data protection and breach response are concerned:

- **GDPR**: Enacted in 2018, GDPR is a comprehensive privacy law that affects any organization handling the personal data of individuals in the **European Union** (EU). GDPR's reach is global – even if your company is not in Europe, if you have EU customers or users, you likely have to comply. How does GDPR tie into cybersecurity? First, it imposes a duty to secure personal data. Article 32 mandates organizations to implement "*appropriate technical and organizational measures*" to ensure a level of security appropriate to the risk (encryption, pseudonymization, resilience, etc.). If an attacker in the *Reconnaissance* or *Exploitation* phase finds you have lax security (say, no encryption or poor access controls), regulators might view that as GDPR non-compliance. Second, GDPR has a breach notification requirement. If personal data is breached (which usually happens in *Actions on Objectives* when data is exfiltrated or accessed by attackers), you must notify your national data protection authority within 72 hours of becoming aware. You also must notify the affected individuals "without undue delay" if the breach is likely to result in high risk to them (for example, if their sensitive information or passwords were stolen). Failing to report can incur penalties separate from the breach itself. GDPR enforcement has teeth: fines can be up to €20 million or 4% of global annual turnover, whichever is higher. We've seen big fines for companies that suffered breaches due to inadequate security (British Airways was fined £20m, Marriott £18m, etc., for breaches). So, under GDPR, each CKC phase has a backdrop: *Recon* – don't expose personal data via misconfigured systems; *Delivery/Exploitation* – have defenses to prevent intrusions; *Installation/C2* – detect and stop unauthorized data access; *Actions On Objectives* – if a breach happens, act swiftly to minimize harm and report. GDPR also promotes **data minimization** – only collect what you need. That helps reduce impact if a breach occurs (less data for attackers to steal).

- **California Consumer Privacy Act/California Privacy Rights Act (CCPA/CPRA)**: California led the U.S. in comprehensive privacy law with CCPA (effective 2020) and its amendment CPRA (most provisions effective 2023). These laws grant California residents rights over their personal information and place obligations on businesses. In a cyber incident context, one unique aspect is that CCPA gives consumers a right to sue (a "private right of action") if certain sensitive personal information is compromised due to a business's failure to implement reasonable security. This is somewhat narrow – it usually applies to breaches of data such as social security numbers, driver's license numbers, financial information, and so on. But it means that after a hack, you could face class-action lawsuits in California if you were deemed negligent with security. The law sets no strict standards, but "reasonable security" is expected (which, in practice, pushes companies to follow frameworks such as CIS Controls, NIST, etc., to show they took precautions). CPRA, which strengthens CCPA, also established the California Privacy Protection Agency, which can issue fines for violations. While CCPA doesn't have a 72-hour rule like GDPR, California has a general breach notification law (as do all 50 U.S. states) requiring notification to individuals typically within 30–45 days, and sometimes to the state Attorney General if a large number of people are affected. So, in the CKC context, after an *Actions on Objectives* data breach, you'd be preparing those notices for Californians and possibly bracing for lawsuits alleging you didn't adequately protect their data. Ensuring you have reasonable cybersecurity measures at each phase (such as access controls, encryption, and patching regimen) not only helps prevent breaches but also provides legal defensibility under CCPA if something does happen.

- **Health Insurance Portability and Accountability Act (HIPAA)**: This is a sector-specific law in the U.S. governing healthcare data, such as **protected health information (PHI)**. If you work in healthcare or handle health records, HIPAA's Security Rule requires safeguarding electronic PHI. That includes controlling access (relevant in *Recon/Exploitation* – ensure only authorized users get in), maintaining audit logs (so you can detect misuse in C2 or Actions phases), and ensuring confidentiality/integrity of data (encrypting data at rest and in transit to mitigate damage even if attackers get in). HIPAA also has a Breach Notification Rule: if PHI is breached, you must notify affected individuals and report to the Department of Health and Human Services (and for larger breaches, notify media as well). The timeline is no later than 60 days from discovery, but prompt notification is encouraged. Penalties for non-compliance can be hefty (up to millions of dollars, scaled by negligence level). For instance, if an attacker exfiltrates patient records (Actions On Objectives phase) and it turns out the organization hadn't encrypted those records or hadn't patched known vulnerabilities, regulators can levy fines for those lapses. So practitioners in healthcare need to align their security controls with HIPAA requirements at each CKC phase – for example, strong authentication to reduce the risk of exploitation via stolen credentials, continuous monitoring to catch intrusions, and contingency plans (backups/disaster recovery) to deal with ransomware events so patient care isn't disrupted.

- **Personal Information Protection Law (PIPL)**: Effective from November 2021, PIPL is China's comprehensive data protection law, akin in many ways to GDPR, but with some differences. If your organization handles the personal data of people in China (perhaps you have users or a subsidiary there), PIPL could apply. It requires consent for data collection, mandates data localization for certain sensitive data (meaning that data should be stored in China), and imposes breach notification duties. Under PIPL, if personal information is leaked, tampered with, or lost, the organization must promptly take remedial actions and notify authorities and affected individuals. There are also provisions about using personal data in automated decision-making and so on, but from a cybersecurity standpoint, PIPL again reinforces that companies must secure personal data and respond to incidents. Notably, China also has a Cybersecurity Law (2017) that sets network security requirements, and regulations for "*critical information infrastructure*" operators with stringent security assessment rules. Chinese authorities have been known to mete out penalties and even detain company officials if a breach is seen as caused by negligence. For instance, after a major hotel chain breach, the company's China executives faced legal trouble. In CKC terms, failing to stop an attack or respond properly in China can have not just fines but personal consequences for executives. So, if operating in that environment, rigorous compliance (including regular security risk assessments mandated by law) is both a legal and moral imperative.

- **Lei Geral de Proteção de Dados (LGPD)**: Brazil's data protection law, effective 2020, is another GDPR-like framework. It similarly requires protecting personal data and has breach notification duties. Organizations must report data breaches to the national authority and sometimes to individuals, although specifics are case by case (the regulation left some flexibility). LGPD can fine up to 2% of a company's Brazilian revenues, capped at 50 million reais per violation. If you have Brazilian customer data, you'd treat a breach much like under GDPR – prompt assessment and notification. Brazilian regulators have begun enforcement, though fines so far have been modest as the regime is new. Still, expect that to ramp up. One nuance is that LGPD, like GDPR, applies broadly, so even a cyber incident that exposes Brazilian personal data via a hacker triggers it, regardless of where the hacker or company is.

- **Other international frameworks**: Many other countries have their own laws (Canada's PIPEDA, Japan's APPI, South Africa's POPIA, etc.), but they share common themes of requiring reasonable security and breach notification. There are also international cooperative frameworks. The Budapest Convention on Cybercrime (2001) is a treaty joined by many countries that harmonizes cybercrime laws and facilitates cooperation. When a cross-border cybercrime happens (which is almost always), this treaty helps, for example, in the CKC C2 phase by enabling law enforcement in one country to legally obtain server logs from another country's servers or to extradite cybercriminals. Knowing that your country is part of such cooperation can assure you that involving law enforcement is worthwhile because they have channels to pursue attackers internationally. Another non-binding but influential framework is the NIST **Cybersecurity Framework (CSF)** – not a law, but a widely recognized set of best practices (identify, protect, detect, respond, and recover) that many regulations reference. Aligning your security program with NIST CSF or ISO 27001 can indirectly help comply with multiple laws since regulators often map their requirements to these standards.

To tie it together, here is a simplified mapping of some major legal frameworks to CKC phases and what they mainly impact:

CKC phase	Relevant legal frameworks and requirements	Key obligations
Reconnaissance	<ul><li>Computer misuse laws (unauthorized scanning)</li><li>Data protection laws (GDPR, etc.) on personal data usage in threat intel</li></ul>	<ul><li>Do not perform unauthorized scanning on others</li><li>Protect any personal data you collect (minimal and lawful basis)</li></ul>
Weaponization	<ul><li>Laws against distribution of hacking tools (UK CMA, etc.)</li><li>Export controls (for exploit tools)</li></ul>	<ul><li>Don't share or sell malware/exploits</li><li>Ensure any security testing tools are used under authorization</li></ul>
Delivery	<ul><li>Anti-hacking laws (CFAA, etc.)</li><li>Communications laws (CAN-SPAM for phishing?)</li><li>Corporate monitoring policies (for email/web filtering)</li></ul>	<ul><li>Attackers violate these, so defenders must ensure that monitoring is disclosed and lawful</li><li>Possibly need user consent for interception in some jurisdictions</li></ul>

Exploitation	• Cybercrime laws (unauthorized access, causing damage) • Industry regulations (**Payment Card Industry Data Security Standard (PCI-DSS)**) if card data accessed)	• Prevent unauthorized access to protected data (due care) • If cardholder data is compromised, follow PCI breach procedures
Installation	• Cybercrime laws (unauthorized implantation of malware) • Lawful access/warrant needed for any defensive "active" measures on attacker systems	• Remove malware from your systems (duty to clean up) • Do not counter-hack • Preserve evidence before cleanup for legal purposes
C2	• Cybercrime laws (misuse of networks) • Privacy laws (when monitoring outbound traffic) • Budapest Convention (law enforcement cooperation for takedowns)	• Monitor your network within legal bounds • Share indicators with authorities or CERTs under info-sharing laws (e.g., the Cybersecurity Information Sharing Act in the U.S. encourages sharing threat information) • Cooperate with cross-border law efforts
Actions on Objectives	• Data breach notification laws (GDPR, CCPA, LGPD, etc.) • Sectoral laws (HIPAA, financial regulations, etc.) • Sanctions law (if paying a ransom to a sanctioned entity) • Negligence liability (common law duty of care, class actions, etc.)	• Notify regulators and affected individuals of breaches within set timeframes • Possibly inform sector regulators • If paying a ransom, ensure you are not violating sanction rules • Post-incident, address any compliance gaps to avoid penalties (fines or lawsuits)

Table 10.1 – Major legal frameworks in different phases in the Cyber Kill Chain

This table isn't exhaustive, but it shows how various laws come into play across the attack lifecycle. For example, early phases are more about not overstepping as a defender and being mindful of computer misuse, whereas later phases invoke breach notification and consumer protection laws.

Having covered the mandates that guide cyber operations worldwide, let's address one of the most polarizing debates in cyber defense: whether defenders can legally or ethically hack back.

Unauthorized hacking back and active defense

One of the most debated legal/ethical topics in cybersecurity is how far a defender can go in striking back at an attacker. The frustration is understandable: when under attack, you might want to *do unto them as they did unto you* – trace the intruder and maybe shut them down or retrieve your stolen data. This concept is often referred to as *hacking back* or *active cyber defense*. Let's clarify definitions. **Passive defense** is traditional defense – firewalls, intrusion prevention, and so on – which stop the attacker at your border. **Active defense** can mean a spectrum of things, some of which are legal (such as creating honeypots or beacon documents that stay within your network but gather intel on the attacker) and some of which cross into hacking back (such as accessing or damaging the attacker's systems). The question is: *Can we ever legally hack the hacker?*

The current answer in most jurisdictions is *No* – it's generally illegal for private entities to engage in offensive actions, even against an attacker. In the U.S., for instance, the CFAA does not have an exception that allows victims to retaliate. If you break into the attacker's computer or even just access it without permission, you become an unauthorized attacker under the law. There have been attempts to change this: a notable proposal was the **Active Cyber Defense Certainty Act** (**ACDC Act**), a bill introduced in Congress a few years ago, which would have given hacked companies a narrow legal right to conduct certain active defense measures (such as tracking the attacker or destroying data the attacker stole, by accessing the attacker's systems). This bill did not become law, amid many concerns. Law enforcement and experts generally opposed it, warning that hack-back could lead to vigilantism, mistaken identity (you might hit the wrong server if attackers bounce through innocent parties), and escalation. The **U.S. Department of Justice** has strongly discouraged hacking back, and organizations risk both civil and criminal liability if they misstep. A similar stance exists in Europe – no country legally permits private hack-back that I'm aware of; self-defense in cyberspace isn't a right the way physical self-defense can be. The only "active defense" that's encouraged is sharing information with authorities or maybe employing deception techniques within your own network.

Ethically, hacking back is also fraught. You could harm innocent bystanders. For example, an attacker may use a compromised university server as a jump point. If you retaliate against that server, you're essentially attacking an innocent party's system, compounding the crime. Even if you do find the real source, you might trigger a bigger conflict – if it's a nation-state actor, hacking back could have international consequences. There's also the question of proportionality and authority: who are we as private citizens or companies to be judge, jury, and executioner in cyberspace? The consensus is that it's best left to law enforcement or the military, who operate under legal frameworks and oversight.

That said, active defense within your own turf is acceptable and often wise. This includes things such as deploying beacon files (documents or data that phone home when opened, alerting you that someone accessed them – staying within legal lines because the attacker essentially triggers it on their machine), setting up honeypot systems to lure attackers (legal on your network; just don't hack the attacker from it), and doing threat attribution analysis (tracking where the attacks are coming from, perhaps even engaging in dialogue with the attacker in a controlled manner). Some companies engage with attackers in negotiation chats not just to potentially resolve the incident but also to gather intel such as cryptocurrency wallets or hacker identities, which they then pass to law enforcement. As long as you're not exploiting the attacker's systems or violating any laws in that engagement, it can be a gray but possibly justifiable area.

Practitioner guidance on hack-back

The best practice is to avoid offensive actions and instead focus on robust defense and involving authorities. If you're interested in the active defense arena, consult legal counsel before implementing anything that toes the line. For example, if you wanted to plant a "GPS beacon" in a file that got stolen so you could locate the thief – is that legal? It might be considered unauthorized access to the thief's device or a violation of wiretap laws, depending on how it works. You'd need legal guidance. In many cases, the safer route is to simulate or predict where stolen data might surface (such as monitoring dark web markets for your data, which is legal) rather than electronically tracking it via a beacon (which could be construed as hacking the thief's machine). Keep in mind also the safety: engaging a hacker could provoke them to react more destructively (if they realize you're fighting back, they might retaliate against your company harder). Unless you're equipped for that fight, don't initiate it.

The advice from experts is *don't hack back. Work with law enforcement, improve your defenses, and use legal means to pursue attackers.* It might not be as viscerally satisfying, but it keeps you on the right side of law and ethics. Notably, many insurance policies and regulations would not cover or may penalize a company that tried to hack back and caused collateral damage. Your goal in incident response is containment and recovery, not vengeance.

To sum up, while the idea of striking back at cybercriminals is alluring and occasionally gets floated in legislatures, as of now, it remains off-limits for practitioners. Stay informed on laws (they could evolve), but proceed very cautiously and always under legal guidance if venturing beyond your network. Meanwhile, channel that energy into better defense and perhaps advocacy for stronger law enforcement action against cybercrime globally.

Ethical leadership and compliance best practices

Building a resilient cybersecurity program isn't just about technology – it's equally about governance, leadership, and culture. In the face of all these legal obligations and ethical dilemmas we've discussed, having strong leadership and a compliance-minded culture is the glue that ensures your organization "*does the right thing*" consistently. Here, we outline some best practices for leaders and teams:

- **Tone from the top**: Leaders, especially CISOs and CIOs but also the CEO and board, need to set the tone that cybersecurity and ethics go hand in hand. This means executives openly prioritize security (not treating it as just an IT problem) and also emphasize ethical conduct (e.g., not hiding bad news and not cutting corners on privacy). A simple example is a CEO supporting the decision to disclose a breach promptly (even if it's embarrassing) because it's the right thing to do for customers. When employees see leaders making ethical choices in security, it filters down.

- **Define policies and codes of conduct**: Develop clear cybersecurity policies that include sections on legal compliance and ethical behavior. This might involve a **code of ethics for IT/security staff** – for example, statements such as "*We respect privacy in carrying out security tasks,*" "*We will only use our access for legitimate business purposes,*" and "*We comply with all applicable laws in investigations and will not exceed our authority.*" Many security professional organizations (such as (ISC)² for CISSPs) have codes of ethics you can draw on. Ensure employees know that violating laws or ethical principles (such as snooping on data without cause) is not acceptable. At the same time, give them guidance on what *is* acceptable (such as reporting a coworker who is circumventing security out of concern, which is ethical even if it feels like tattling – you are protecting the organization).

- **Training and awareness**: Regularly train not just general employees on security awareness but also the security team on relevant legal regulations and ethics. Your security analysts and engineers should have at least a basic understanding of laws such as GDPR, CCPA, and so on, so they recognize when something might have legal implications. For example, if an analyst discovers personal data in log files that shouldn't be there, they can flag it as a privacy concern. Provide scenario-based ethics training too: for example, walk through a hypothetical breach and ask, "*Would we tell customers about this detail or keep it quiet? What's the right thing to do?*" This preps the mindset to default to transparency and integrity.

- **Privacy and security collaboration**: In many organizations, a chief privacy officer or data privacy team exists alongside the security team. Close collaboration between these functions is critical. Together, they can conduct **data protection impact assessments (DPIAs)** for new technologies to ensure security measures are balanced with privacy. For instance, if deploying user monitoring, the privacy team can help minimize data collection or anonymize logs. This partnership ensures compliance with data laws and aligns with ethical data handling. It's also increasingly common for companies to have **privacy by design** and **security by design** principles – meaning any new project must build in compliance and security from the start, not as an afterthought.

- **Incident response readiness**: We have talked about **incident response (IR)** a lot. From a leadership view, ensure an IR plan is not just on paper but tested. Do tabletop exercises with executives: simulate a ransomware scenario and have legal and PR involved in the mock drill. This way, if the real thing hits, the leaders have muscle memory for making ethical decisions under pressure (e.g., mock-debating *"Do we pay or not?"* in an exercise gives insight into how to approach the real thing). After tests or real incidents, do after-action reviews and actually implement the recommended improvements – this follow-through is key and shows the organization's commitment to getting better (which is part of an ethical culture of continuous improvement).

- **Compliance management**: Given the array of laws (some covered previously), larger organizations may have a compliance officer or team tracking all this. If you don't, it's wise to at least maintain a compliance matrix – a document or tool that lists each relevant law/regulation, what it requires for cybersecurity, and your status (compliant, needs work, or not applicable). For example, you might list GDPR, and note *"Have breach response plan covering 72-hr notification; need to improve encryption of laptops to meet GDPR Article 32."* Do similar for HIPAA and so on, depending on your business. Regularly review and update this matrix. It will help prioritize projects (if a new state privacy law comes out, you add it and then adapt your processes accordingly).

- **Third-party risk and compliance**: Your partners and vendors can pose legal/ethical risks too. If you outsource data processing, ensure those vendors contractually commit to security and privacy standards (DPIAs for GDPR, business associate agreements for HIPAA, etc.). Perform due diligence on their security practices. If they suffer a breach, you often still hold a responsibility to report and deal with the fallout. A classic example is the MOVEit breach in 2023 – many companies had to notify on behalf of that vendor's failure. So, managing supplier risk is both a legal necessity and an ethical one (you want your customers' data safe even when it's with a vendor).

- **Proportional security (avoid security theater)**: Ethically, we have to justify our security measures. Surveillance or heavy-handed controls that aren't actually effective or necessary can erode trust without real benefit. For instance, recording every employee's screen all day "just in case" is overkill and likely violates privacy principles. Instead, implement controls proportional to risk. If an employee works with very sensitive data, maybe you monitor access to that data, but you don't need their webcam feed! Ensuring your measures are justified and documented as such also helps legally (if an employee ever questions a privacy intrusion, you can show the rationale and that it aligns with law and policy).

- **Whistleblower protection**: Encourage a culture where, if someone sees an unethical practice, they feel safe to report it. Say a developer finds that a product is logging user passwords in plain text (yikes!). They should be able to bring that up without fear. Have channels (even anonymous hotlines) to report security or privacy concerns. Legally, some laws protect whistleblowers, but beyond law, it's about culture. Ethical leadership thanks employees who point out vulnerabilities or compliance gaps; it doesn't punish them. This also means not shooting the messenger if a security researcher externally reports a bug – treat them with respect (assuming they follow responsible disclosure). Companies that threaten researchers or hide issues end up worse off.

- **Continuous improvement and accountability**: Finally, compliance and ethics are not checkboxes; they are ongoing. Conduct periodic audits (internally or hire outside assessors) to verify that what you think is happening (per policy) is actually happening in practice. If they find gaps, fix them. Share results with senior management to keep them in the loop – this accountability ensures everyone knows we're serious about both complying with the law and upholding ethical standards. Many organizations find value in getting certifications (such as ISO 27001 or SOC 2) to validate their practices; while those aren't purely about ethics, they instill discipline that supports legal compliance and trustworthiness.

In essence, ethical leadership in cybersecurity fosters a security environment where doing the right thing is the norm, not the exception. It aligns the organization's values with its security mission, ensuring that not only are we guarding data and systems but we're also honoring the people behind that data – our customers, employees, and community. By adhering to best practices and nurturing a culture of integrity, cybersecurity professionals can be confident that they are on the right side of both the law and moral responsibility when navigating the CKC.

Case studies: real-world lessons in legal and ethical consequences

Let's examine a few high-profile cyber incidents with an eye toward the legal and ethical outcomes we've discussed. These cases illustrate what can go right or wrong and provide concrete lessons for practitioners:

Case study 1: the SolarWinds supply chain breach (2020)

In 2020, attackers (now attributed to Russia's foreign intelligence service) conducted an extremely sophisticated supply chain attack. They breached SolarWinds, a maker of network management software, and inserted malicious code into a routine software update of the SolarWinds Orion product. When SolarWinds unwittingly shipped this tainted update to customers, it created a backdoor (dubbed "Sunburst") on the customers' systems. Around 18,000 organizations, including U.S. government agencies and many Fortune 500 companies, received the compromised update, though the attackers ultimately activated the second stage of the attack on a smaller subset of high-value targets (such as the Department of Justice, Microsoft, FireEye, etc.). The breach was discovered when FireEye (a security company also using Orion) found unusual activity and eventually traced it to the Orion software.

Legal consequences

This was a watershed supply chain incident. SolarWinds as a company faced intense scrutiny. While they were a victim of nation-state espionage, questions arose about their security practices (for instance, it became public that, at one point, they had a weak password, `solarwinds123`, publicly accessible – though not necessarily the entry point for this attack, it hurt their image). Investors filed lawsuits claiming SolarWinds failed to disclose security risks, and there were regulatory investigations.

Government agencies impacted had to report to Congress on the extent of damage. This case also led to legal and policy change: the U.S. government issued an executive order on cybersecurity in 2021 that, among many things, required federal software suppliers to adhere to stricter security, such as providing a **software bill of materials** (**SBOM**). From a cross-border legal standpoint, this attack violated numerous laws (unauthorized access, etc.) but since it was likely carried out by Russian intelligence officers, there's little chance of prosecution – instead, the response was diplomatic (sanctions against Russia and expulsions of diplomats). The incident really underscored the need for international norms on not targeting supply chains – an issue now discussed in diplomatic channels.

For practitioners, SolarWinds highlights the duty of care in software development and updates. Legally, if you supply software, you could be liable for downstream effects if you were negligent in security. Ethically, your compromise can become everyone's compromise – so securing build systems, code signing, internal access, all of that is paramount. SolarWinds claims to have significantly improved their security since then (they had to in order to regain trust), implementing things such as expanded monitoring, **multi-factor authentication** (**MFA**) everywhere, and external audits. Another takeaway is the importance of transparency. SolarWinds and FireEye were relatively transparent once this was discovered. FireEye even publicly shared detection rules and countermeasures, which was an ethical action and helped countless others check whether they were breached. SolarWinds cooperated with investigators and communicated updates to customers. Ethically and legally, this cooperation was essential to help victims mitigate damage.

Lessons learned

Supply chain attacks mean that even if your kill chain looks solid, an attacker may piggyback through a trusted update – so zero trust principles and monitoring are key. Legally, expect regulators to ask whether you vetted your suppliers. In the SolarWinds aftermath, companies began demanding SBOMs from vendors and doing more rigorous third-party risk assessments. Also, the need for international collaboration was clear: no single country could unravel or respond to this alone. Information sharing through channels such as CERTs was critical. Practitioners should engage in communities, such as **information sharing and analysis centers** (**ISACs**) for their sector, to share and receive such intel swiftly.

Case study 2: the Colonial Pipeline ransomware attack (2021)

In May 2021, Colonial Pipeline, which operates a major fuel pipeline on the U.S. East Coast, fell victim to ransomware (by the criminal group DarkSide). The attack led Colonial to shut down pipeline operations as a precaution, causing fuel shortages and panic buying in several states. This was one of the most disruptive cyberattacks on U.S. critical infrastructure to date. Colonial decided to pay the ransom (approx. $4.4 million in Bitcoin) to obtain the decryption tool and restore operations quickly. The FBI was involved early and later managed to recover a portion of the ransom by tracking the Bitcoin wallet (seizing about $2.3 million of it) – a rare bit of good news in a ransomware scenario.

Legal consequences

This incident rang alarm bells \at the highest government levels. Legally, it exposed that the pipeline industry wasn't under as strict cybersecurity regulations as, say, the power sector. Following the attack, the U.S. **Transportation Security Administration** (**TSA**) issued security directives mandating pipeline companies to implement specific cyber measures and report incidents. So, a direct outcome was new regulatory requirements – effectively law – to raise the cybersecurity bar for pipelines. Colonial's decision to pay the ransom was legally permissible (DarkSide was not at that time sanctioned, and paying a ransom is not illegal). However, it raised ethical questions and policy discussions on whether paying ransoms should be banned or discouraged more strongly. The attackers themselves had a quasi-ethical stance; DarkSide issued a strange apology for the societal impact, stating they only wanted money, not to cause problems – highlighting an odd dynamic where even criminals felt the need to justify themselves publicly. Colonial's CEO testified before Congress, facing tough questions about their preparedness. One legal sticking point was that their outdated VPN system didn't have MFA, which was the likely entry point via a compromised password. While not illegal to lack MFA, it was certainly seen as a cybersecurity lapse. The court of public opinion and regulators made an example that critical operators must adopt basic security hygiene.

For practitioners, Colonial Pipeline exemplifies the *proportionality and public safety* part of cybersecurity ethics. Shutting down the pipeline was a drastic measure that had a huge public impact, but was arguably the right call to ensure the malware didn't spread to the more sensitive operational technology networks controlling the pipeline flow. Legally, once the pipeline shutdown led to economic effects, the government stepped in heavily – including emergency declarations for fuel transport. This shows that a cyber incident can quickly become not just an IT issue or even a corporate issue but a national issue, invoking legal powers far beyond your control. It underscores the importance of\ IR planning that considers public communication and coordination with the government. Colonial was criticized for not having a specific playbook for such a scenario.

Lessons learned

Ransomware is a scourge that needs both preventive and reactive plans. Make sure you have offline backups, practice restoration, and have contact information for law enforcement handy. Legally, know that if you're a critical infrastructure, regulators will expect reports as soon as possible (Colonial notified certain federal agencies within a day or two due to the impact). Also, this case encouraged more information-sharing; within days, the CISA and FBI published details about the attack indicators, and industry groups convened to discuss defenses. Practitioners should engage in these information-sharing efforts; if you're in a regulated sector, join your sector's coordination council or similar. Ethically, Colonial Pipeline illustrated the tough call on ransoms: they chose to pay for the greater good of restoring the fuel supply. Companies must weigh the ethical implications of either choice and, ideally, that calculus should be thought through *before* an attack (in policy form) so you're not scrambling to decide in the heat of the moment.

Case study 3: the MOVEit data breach (2023)

MOVEit Transfer is a widely used file transfer application by Progress Software. In May 2023, a zero-day vulnerability in MOVEit was exploited by the Cl0p ransomware group. The flaw allowed attackers to steal data from any MOVEit server reachable over the internet. This turned into a mass breach affecting organizations worldwide – banks, universities, government agencies, tech companies… you name it. The attackers stole data from over 2,700 organizations, impacting nearly 95 million individuals by some estimates. Cl0p's motive was data extortion; they didn't encrypt files on the victims' systems, they "only" stole data and then attempted to extort each victim organization by threatening to publish the information on their leak site if not paid. Many organizations did not pay, and consequently, personal data (such as names, addresses, social security numbers, and health data in some cases) was leaked publicly. This breach cascaded through supply chains: for example, a payroll company using MOVEit got breached, leading to dozens of its clients (and their employees' data) being compromised.

Legal consequences

This incident triggered *global breach notifications* on a large scale. Under GDPR, state laws in the U.S., PIPL in China if any Chinese data was involved, and so on, companies had to notify their regulators and affected individuals. For instance, British Airways and the BBC were impacted through a third-party service provider – they notified thousands of employees that their data was stolen. In the U.S., several state Attorneys General opened investigations into companies that had large numbers of affected residents. There was one unique aspect: since the vulnerability was in third-party software, some victim organizations might try to take legal action against Progress Software (MOVEit's vendor) for damages. If Progress was found negligent (say, they ignored reports of the bug or had very weak security development practices), they could be liable. However, suing software vendors for breaches is still a gray area and is often limited by contract terms that disclaim such liability. It will be interesting to see whether any class actions succeed there. Regulators, at minimum, will look at whether companies patched swiftly once the vulnerability became known and whether they had proper vendor risk management. Progress did release a patch quickly (within days of discovery), so the blame is tempered by that fact – you can't realistically expect zero vulnerabilities ever; the key is the response. Legally, this case has pushed even more focus on software security accountability. The U.S. Cyber Safety Review Board (a government body) announced an investigation into the MOVEit event, which could result in recommendations or policy changes around secure development and vulnerability management.

Ethical consequences

For many affected companies, an ethical challenge was how transparent they should be about a breach that wasn't directly their fault but hit their customers' data. Most took the high road and promptly informed their customers, rather than hiding behind the "*it was our vendor's fault*" excuse. Ethically, that's the right approach – customers entrusted their data to you, not to your vendor, so you should own up to the responsibility of informing them and helping them mitigate any harm. Here is another ethical angle: Cl0p gave some organizations an ultimatum not to disclose the breach or they'd publish data immediately (trying to muzzle victims). Notifying authorities is a legal must in many cases, so victims had to ethically stand up to that threat and still do the right thing by notifying, even if it meant data might get leaked sooner. Also, this saga showed how a lack of ethical restraint by one software vendor (one might argue perhaps Progress could have found the bug earlier with thorough code audits) can inadvertently harm millions of people – though again, hindsight is 20/20, and no major indication that Progress was negligent; it was just an unforeseen flaw that got weaponized.

Lessons learned

Third-party software can be an Achilles' heel – always stay on top of patches for them and isolate such applications if possible (many now isolate file transfer apps in segmented networks). When a zero-day is announced, treat it as an emergency – Progress gave patch instructions; some who applied immediately were able to stop breaches, and those who delayed even a day or two often got hit, given the speed of mass exploitation. From a compliance view, ensure contracts with vendors include timely vulnerability disclosure and patch obligations. Many organizations are now demanding that vendors have secure development certifications or pen-test their products regularly. The MOVEit case also underscores the importance of having a broad incident response net – lots of organizations didn't realize they were breached until the hacker's extortion email or they saw their name on Cl0p's leak site. Investing in capabilities to detect anomalous file downloads or sudden activity from service accounts could help catch such breaches earlier. Lastly, it's a reminder that you can do everything right in your kill chain, but an attack on a common software used by many can still snag you – hence why community defense (sharing information quickly about attacks) and pushing for higher software security standards collectively is crucial.

Each of these cases reinforces elements of what we've covered: SolarWinds for supply chain diligence and international cooperation; Colonial for critical infrastructure responsibility and the ransom decision; and MOVEit for third-party risk and the need for swift transparency. By studying them, we internalize those lessons and can apply them to our own strategies, hopefully avoiding being the next cautionary tale.

Practitioner recommendations and checklists

Bringing it all together, here are some practical recommendations and a checklist to help cybersecurity professionals integrate legal and ethical best practices into their defense:

- **Integrate legal compliance into security processes**: Don't treat legal/compliance as a separate silo. For each security control or process, ask "*Are we meeting our legal obligations here?*". For example, when setting up log management, consider data retention laws and privacy (don't keep personal data logs longer than needed). When designing IR, bake in breach notification steps per the laws that apply to you. Maintain a calendar of key regulatory requirements (security policy reviews, user training, and third-party assessments) so nothing falls through the cracks.

- **Maintain an up-to-date "security laws and regulations" knowledge base**: As part of your team documentation or wiki, keep summaries of relevant laws (GDPR, etc.) and what they require in layman's terms for your technical staff. Translate legal mandates into actionable security controls. For example, GDPR requires the ability to ensure ongoing confidentiality – map that to controls such as encryption, access management, and disaster recovery drills.

- **Foster a culture of ethics**: Encourage team members to voice concerns if they feel something is unethical or legally questionable. Create avenues for discussion – maybe a monthly meeting where the team discusses a news incident's ethical aspects, or an internal anonymous Q&A. Ensure that when tough decisions arise (such as whether to monitor something invasive), the team weighs the ethical pros/cons, not just the technical outcome.

- **Plan for the worst (and learn from it)**: Have a robust IR plan that covers scenarios such as data breaches, ransomware, and insider misuse. Include in the plan the steps for internal ethics review and decision-making (e.g., convening an ethics panel if considering paying a ransom or dealing with sensitive breach communications). After any incident or even drills, do post-mortems that include, *"Did we act in accordance with our values and legal duties? What could we improve?"*.

- **Engage with peers and authorities**: Build relationships with local law enforcement cyber units and industry peers. Knowing the FBI field office contact or your country's cybercrime unit contact can save precious time during an incident (you won't hesitate to call if you know who's on the other end). Join information-sharing groups (such as an ISAC). This not only helps you receive early warnings (such as C2 indicators and vulnerability alerts) but also shows regulators you're proactive (some regulators ask whether you participate in such exchanges as a positive factor).

- **Use ethical decision frameworks**: When confronted with a gray area, use frameworks such as utilitarian (what action results in the most good or the least harm), deontological (what is our duty or principle, such as honesty), and virtue ethics (what aligns with our company values). For instance, in breach communication, *utilitarian* might argue to delay notifying to avoid panic, but *deontological* would say we have a duty to inform transparently. By examining both, you can reach a balanced decision.

- **Regular audits and assessments**: Periodically audit not just technical controls but also compliance with policies and the effectiveness of training. Have an external party or internal audit check, for example, whether all employees have completed security and privacy training, whether your vendors all signed the updated data protection addendum, whether your honeypot is configured not to gather excessive data, and so on. Treat these as learning exercises, not punitive.

- **Keep documentation**: During or after a crisis, good documentation can show you followed due process. Keep incident logs of decisions (with rationale). Keep copies of communications to authorities. Document your risk assessments for why certain security measures are or are not in place. This can be your legal savior or at least a mitigating factor if regulators come knocking – it shows you were conscientious and systematic.

Finally, here's a concise *Legal and Ethical Readiness Checklist* you can use:

- **Clear authorization** for all security activities (penetration tests, monitoring, etc.). Do we have policies/agreements in place?

- **Privacy considerations** assessed for security tools (DLP, EDR, etc.). Have we done a DPIA for our monitoring?

- **Employee training** on security policies, phishing, and also on ethics (such as data handling, reporting issues, etc.).

- **IR plan** that includes legal notification steps and an ethics review step for major decisions.

- **Contact list** of legal counsel (including privacy officer), law enforcement, and IR vendors at the ready.

- **Breach notification templates** pre-drafted (so we're not writing from scratch under duress).

- **Tabletop exercises** conducted in the last 12 months, including one with an ethical dilemma (e.g., a ransom scenario).

- **Compliance register** updated for all laws/regulations that apply, with assigned owners for each requirement.

- **Third-party security reviews** current (we've assessed critical suppliers for their security, and they contractually commit to our standards).

- **Cyber insurance** (if we have it) understanding – including what they require us to do (notifications, use of certain vendors, etc.) so we don't void coverage.

- **A "no hacking back" policy** explicitly stated and communicated to the team.

- **Up-to-date system diagrams and data flow maps** (helps in quickly assessing what data might be affected during a breach – important for legal notice).

- **Encryption and access control** in place for sensitive data (at rest and in transit) – often a legal safe harbor (many laws don't require notification if data was encrypted and unusable to thieves).

- **Regular patches and vulnerability management** documented – showing we patch critical vulnerabilities within X days (compliance with "reasonable security").

- **Ethical leadership engagement** – such as a quarterly cybersecurity brief to the board that includes not just status but also ethical considerations and decisions made.

Following this checklist won't guarantee you never have an incident or legal issue, but it puts you in a defensible and principled position. You can show that you were proactive, compliant, and acting in good faith, which significantly reduces legal liability and preserves trust with customers and stakeholders even if something does go wrong.

Summary

This chapter emphasized the critical intersection of cybersecurity practices with legal regulations and ethical standards. It underscored that effective cybersecurity is not solely about technical defense mechanisms but also involves navigating complex legal requirements and maintaining ethical conduct. Each phase of the CKC presents distinct legal and ethical challenges, from the legality of scanning and monitoring to the ethical considerations of simulated attacks and employee testing. The chapter also explored how global cybersecurity laws such as GDPR, CCPA/CPRA, HIPAA, PIPL, and LGPD impose obligations on organizations to implement robust security measures, notify affected individuals in case of data breaches, and adhere to strict data protection protocols.

The chapter provided practical insights into key legal implications for each CKC phase, such as the restrictions on unauthorized reconnaissance, the ethical boundaries of red teaming, and the obligations to report data breaches under various data protection laws. It highlighted real-world case studies such as SolarWinds, MOVEit, and Colonial Pipeline, illustrating how failures to adhere to legal and ethical guidelines can result in severe consequences, including regulatory fines, legal actions, and reputational damage. Furthermore, the discussion extended to emerging topics such as "hacking back," the ethical use of honeypots, and the duty of care in cybersecurity operations. Practitioners were guided on how to align their security measures with legal frameworks while maintaining ethical conduct in adversarial simulations, incident response, and communication with affected stakeholders.

Additionally, the chapter addressed the broader implications of international cybersecurity laws and the increasing convergence of data protection regulations worldwide. By mapping each CKC phase to specific legal frameworks, the chapter provided a structured approach for organizations to understand their legal responsibilities and the ethical considerations at every stage of a cyber incident. It emphasized the importance of documentation, transparency, and stakeholder communication, especially during critical phases such as *Installation* and *Actions on Objectives*, where decisions around ransomware payments and data breach disclosures can significantly impact both legal outcomes and public perception. Ultimately, the chapter served as a comprehensive guide for cybersecurity professionals to navigate the intricate landscape of legal and ethical obligations, reinforcing the principle that robust security practices must be both technically sound and legally defensible.

In the next chapter, we shift our focus to the proactive defense strategies essential for staying ahead in the rapidly evolving cyber threat landscape. By examining advanced threat intelligence, adaptive response mechanisms, and continuous monitoring, we will outline how organizations can not only detect but also anticipate and neutralize sophisticated cyberattacks. This chapter aims to equip you with actionable frameworks and insights to fortify your cybersecurity posture against emerging threats.

Further reading

The following are some bonus reading materials for you to check out:

- *Budapest Convention on Cybercrime (Council of Europe)*: An international treaty that coordinates laws against cybercrime among member states. It clarifies what constitutes illegal online activity and streamlines cross-border investigations (`https://www.coe.int/en/web/cybercrime/the-budapest-convention`).

- *GDPR (EU General Data Protection Regulation) Text and Guides*: Official information and articles related to Europe's strict data privacy framework. It explains how companies must protect personal data and notify authorities in the event of a breach (`https://gdpr-info.eu`).

- *CCPA/CPRA (California Consumer Privacy Act & California Privacy Rights Act)*: California's main consumer privacy laws, granting residents rights over their personal data. These measures also impose responsibilities on businesses collecting or selling such data (`https://oag.ca.gov/privacy/ccpa`).

- *HIPAA Security Rule (U.S. Department of Health & Human Services)*: Federal guidelines for safeguarding electronic health information. It spells out security measures required of healthcare providers and their partners in the U.S. (`https://www.hhs.gov/hipaa/for-professionals/security/index.html`).

- *PIPL (Personal Information Protection Law of China)*: China's broad data protection legislation covering consent, data localization, and breach notification for organizations handling the personal data of Chinese residents (`https://digichina.stanford.edu/work/translation-personal-information-protection-law-of-the-peoples-republic-of-china-effective-nov-1-2021/`).

- *LGPD (Brazil's General Data Protection Law)*: Brazil's data protection statute, influenced by GDPR. It outlines how companies must handle, store, and process personal data belonging to Brazilian citizens (`https://lgpd.brazilianprivacy.org/`).

- *NIST Cybersecurity Framework (CSF)*: A voluntary framework from the National Institute of Standards and Technology. It helps organizations identify, protect, detect, respond, and recover from cyber threats (`https://www.nist.gov/cyberframework`).

- *Active Cyber Defense Certainty Act (ACDC) Proposal*: A past U.S. legislative proposal exploring a limited right for companies to go on the offensive against hackers, a concept widely debated and never enacted into law (`https://www.congress.gov/bill/115th-congress/house-bill/4036`).

- *Emotet/TrickBot/Botnet Takedown Case Studies (CISA and FBI joint advisories)*: Real-world examples of large botnets being neutralized through international law enforcement action. They illustrate the power of collaboration in dismantling global cybercriminal infrastructure (`https://www.cisa.gov/news-events/cybersecurity-advisories`).

- *SolarWinds Supply Chain Attack*: Analysis from CISA describing how malicious code injected into the SolarWinds Orion update infected thousands of organizations, including high-profile government agencies (`https://www.cisa.gov/uscert/ncas/alerts/aa20-352a`).

- *Colonial Pipeline Ransomware Attack*: U.S. Department of Justice update detailing how the FBI recovered part of the ransom payment made by Colonial Pipeline, highlighting both technical and legal facets of responding to ransomware (`https://www.justice.gov/archives/opa/pr/department-justice-seizes-23-million-cryptocurrency-paid-ransomware-extortionists-darkside`).

- *MOVEit Vulnerability and Exploits*: Progress Software advisories explaining the zero-day flaw in the MOVEit Transfer application and offering patches and guidance to prevent mass data theft (`https://www.progress.com/security/moveit`).

- *Maersk NotPetya Attack Retrospective*: Various post-incident analyses compiled by the World Economic Forum highlight how the NotPetya malware outbreak severely impacted shipping giant Maersk, offering lessons on crisis response and resilience `https://www.weforum.org/publications/cyber-resilience-in-the-oil-and-gas-industry-playbook-for-boards-and-corporate-officers/`).

- *ISO/IEC 27001 Standard*: An international standard specifying how to establish, implement, and continually improve an information security management system. Often used to demonstrate robust cybersecurity practices and compliance (`https://www.iso.org/isoiec-27001-information-security.html`).

Get This Book's PDF Version and Exclusive Extras

UNLOCK NOW

Scan the QR code (or go to `packtpub.com/unlock`). Search for this book by name, confirm the edition, and then follow the steps on the page.

Note: Keep your invoice handly. Purchase made directly from Packt don't require one.

11
The Future

The cybersecurity landscape is on the cusp of a profound transformation, driven by the convergence of cutting-edge technologies, evolving threat landscapes, and the relentless pursuit of robust digital defense mechanisms. As we venture into this new era, the integration of **Artificial Intelligence (AI)** and **Machine Learning (ML)** stands at the forefront, promising to revolutionize our approach to threat detection, response automation, and predictive security measures. While offering unprecedented capabilities in safeguarding digital assets, these advancements also herald the dawn of a new arms race between defenders and increasingly sophisticated adversaries. Alongside AI and ML, the looming advent of quantum computing presents both a formidable challenge to current cryptographic standards and an opportunity for groundbreaking security innovations. The cybersecurity paradigm is further evolving with the widespread adoption of Zero Trust architecture, a model that fundamentally reimagines network security by operating on the principle of continuous verification.

In this chapter, we will explore how these technological advancements and methodological shifts are reshaping the cybersecurity landscape and revolutionizing investigative frameworks such as the Cyber Kill Chain. Developed by Lockheed Martin, the Cyber Kill Chain has long served as a cornerstone for understanding and combating cyberattacks. However, the model must evolve to remain relevant and practical as we stand on the brink of this new era. We will examine how AI-driven analytics can enhance each stage of the Cyber Kill Chain, from detecting subtle reconnaissance activities to swiftly neutralizing advanced persistent threats. Furthermore, we will delve into the implications of quantum computing on offensive and defensive cyber operations and how it necessitates a fundamental reevaluation of our encryption standards and practices.

As we navigate through these transformative developments, it's crucial to consider the broader implications for organizations, cybersecurity professionals, and society at large. The future of cybersecurity demands not only technological prowess but also an adaptive mindset, capable of anticipating and outmaneuvering threats in an increasingly complex digital ecosystem. This chapter provides a comprehensive overview of the challenges and opportunities that lie ahead, equipping you with the insights necessary to navigate the evolving cybersecurity landscape and leverage the Cyber Kill Chain model in this new technological frontier.

Emerging threats in cybersecurity – navigating the future landscape

As we hurtle into the digital future, the world of cyber threats is evolving at breakneck speed. Imagine a landscape where your computer's worst enemy isn't just a hacker in a hoodie but an AI that thinks faster than any human. Welcome to the future of cyber-attacks.

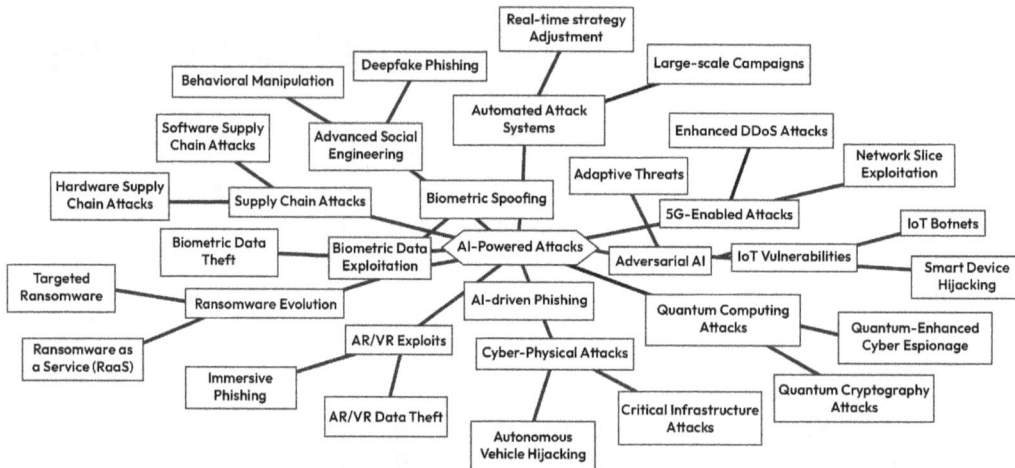

Figure 11.1 – AI-driven defense system

This section explores these emerging threats and provides advanced countermeasures to mitigate them.

AI-powered attacks

AI is poised to revolutionize the cyber threat landscape. Adversarial AI involves using AI algorithms to create more sophisticated and adaptive threats. These AI-driven attacks can learn from and adapt to the defenses they encounter, making them particularly challenging to mitigate. For instance, an AI could generate phishing emails that are highly personalized and convincing based on social media activity and other data sources. Additionally, cybercriminals will leverage AI to automate their attacks, allowing them to launch large-scale, efficient, and highly targeted campaigns. These systems can identify vulnerabilities, exploit them, and adjust strategies in real time based on the success or failure of initial attempts. Such automation will significantly increase the speed and scale of attacks.

Here are some countermeasures:

- **AI-driven defense systems**: Employ AI to develop adaptive defense mechanisms to learn from and counter AI-driven attacks.

- **Behavioral analytics**: Use AI-powered behavioral analytics to detect anomalies and unusual patterns in network traffic.

- **Automated threat response**: Implement AI systems that automatically respond to threats, mitigating real-time damage.

Quantum computing attacks

Quantum computing represents a dual-edged sword in cybersecurity. While it promises unparalleled computational power, it also threatens to undermine current encryption standards. Quantum computers can break traditional cryptographic algorithms, such as **Rivest-Shamir-Adleman (RSA)**, and **Elliptic Curve Cryptography (ECC)**, which rely on the difficulty of factoring large numbers or solving discrete logarithm problems. This breakthrough could render current encryption methods obsolete, exposing vast amounts of sensitive data. Moreover, the immense processing power of quantum computers will enable attackers to sift through massive datasets at unprecedented speeds, facilitating more effective and comprehensive cyber espionage campaigns.

Here are some countermeasures:

- **Post-quantum cryptography**: Develop and implement quantum-resistant cryptographic algorithms to protect data from quantum attacks.

- **Quantum Key Distribution (QKD)**: Use QKD to ensure secure key exchange inherently resistant to quantum interception.

- **Regular cryptographic updates**: Continuously update cryptographic protocols to the latest standards to stay ahead of potential threats. Stronger mechanisms and regular cryptographic updates buy time in the early phase of quantum supremacy by enhancing security and delaying the impact of quantum attacks. They reduce risk exposure and help protect current and historical data against future quantum decryption. This proactive approach keeps defenses ahead of evolving threats.

The **National Institute of Standards and Technology (NIST)** released the first three post-quantum encryption standards on August 13, 2024. These standards are part of an eight-year effort to develop encryption algorithms capable of resisting cyberattacks from quantum computers, which pose a significant threat to current encryption methods. The finalized standards include algorithms designed for general encryption and digital signatures, ensuring the security of a wide range of electronic information, from confidential communications to critical national data. NIST urges computer system administrators to transition to these new standards promptly to safeguard against future quantum threats.

Internet of Things vulnerabilities

The proliferation of **Internet of Things (IoT)** devices, from smart home gadgets to industrial sensors, presents new opportunities for attackers to exploit vulnerabilities. Poorly secured IoT devices can be compromised and enlisted into botnets, networks of hijacked devices that launch coordinated attacks, such as **Distributed Denial of Service (DDoS)** attacks. These IoT botnets can disrupt internet services, financial systems, and critical infrastructure. As more devices become interconnected, attackers will also target smart devices for hijacking, which includes everything from smart home systems to medical implants and industrial control systems. Compromising these devices can lead to physical harm, theft of sensitive personal data, or disruption of essential services.

Here are some countermeasures:

- **Comprehensive device management**: Deploy solutions to monitor and control all IoT devices from a central dashboard.

- **Network isolation**: Keep IoT devices separate from critical systems to prevent cross-network attacks.

- **Firmware updates**: Regularly update the firmware of IoT devices to patch known vulnerabilities.

Biometric data exploitation

As biometric authentication methods such as fingerprint scanning, facial recognition, and voice recognition become more widespread, they will become prime targets for cybercriminals. Attackers will develop techniques to spoof biometric data, allowing them to gain unauthorized access to secure systems. For example, sophisticated 3D printing techniques could create fake fingerprints, while AI-generated deepfakes could mimic voices and faces accurately. The increasing use of biometric data for security purposes also raises concerns about the theft of such data. Unlike passwords, biometric data cannot be changed if compromised. Stolen biometric information can be used for identity theft or sold on the dark web, creating long-term security risks for individuals.

Here are some countermeasures:

- **Multimodal biometrics**: Combine multiple biometric factors (e.g., fingerprint and facial recognition) to enhance security.

- **Liveness detection**: Implement advanced techniques to differentiate between real and fake biometric data.

- **Encryption and secure storage**: Encrypt biometric data and store it securely to protect against theft and unauthorized access.

Some experts argue that biometrics, while secure, should not be used as a standalone authentication method due to potential vulnerabilities. They advocate for its inclusion in **multi-factor authentication (MFA)** to strengthen security by combining it with other factors such as passwords or tokens.

Supply chain attacks

Attackers increasingly target supply chains by embedding malicious code into legitimate software updates or third-party components. This approach can compromise many systems through trusted software vendors. For example, a popular software library used by numerous applications could be compromised, spreading malware to all dependent systems. Beyond software, attackers may also exploit vulnerabilities in the hardware supply chain. Malicious components could be inserted into devices during manufacturing, leading to compromised hardware that can infiltrate networks once deployed. These attacks are particularly insidious as they are challenging to detect and can have widespread impact.

Here are some countermeasures:

- **Third-party risk management**: Regularly audit suppliers' security practices to ensure they meet stringent security standards.

- **Secure software development**: Integrate security into every stage of the software development life cycle to minimize vulnerabilities.

- **Hardware verification**: Implement thorough verification processes for hardware components to detect and mitigate potential compromises.

Advanced social engineering

Using deepfake technology to create compelling audio and video content will revolutionize phishing attacks. Cybercriminals can use deepfakes to impersonate trusted individuals, making it easier to deceive targets into revealing sensitive information or transferring funds. For example, an attacker could create a deepfake video of a CEO instructing an employee to transfer money to a fraudulent account. AI and big data analytics will allow attackers to understand and predict human behavior accurately. Cybercriminals can craft highly personalized and convincing social engineering attacks by analyzing social media activity, purchasing habits, and other data. This level of manipulation can lead to more successful phishing attempts and different types of fraud.

Here are some countermeasures:

- **Continuous employee training**: Regular, engaging security awareness training is crucial to prepare employees for sophisticated social engineering attacks.

- **Contextual authentication**: Implement systems that consider the context (e.g., location and time of access) when authenticating users to detect anomalies.

- **Advanced detection tools**: AI-based tools detect deepfake content and flag suspicious communications.

Ransomware evolution

Future ransomware attacks will be more targeted, focusing on high-value individuals and organizations. Attackers will conduct detailed reconnaissance to craft ransomware that maximizes payment likelihood. For instance, they may target hospitals, financial institutions, or critical infrastructure, where the impact of a ransomware attack can be devastating and prompt a quicker payment response. The availability of **Ransomware as a Service (RaaS)** on the dark web will continue to grow, enabling even low-skilled attackers to launch sophisticated ransomware attacks. RaaS platforms provide all the tools needed to deploy ransomware, including malware development, distribution mechanisms, and payment processing, significantly lowering the barrier to entry for cybercriminals.

Here are some countermeasures:

- **Robust backup strategy**: Maintain secure, offline backups to ensure data can be restored without paying ransoms.

- **Endpoint Detection and Response (EDR)**: Deploy EDR solutions to detect and respond to ransomware threats early.

- **Incident response planning**: Develop and regularly test an incident response plan for ransomware scenarios to ensure quick and effective action.

Cyber-physical attacks

Future attacks will increasingly target essential infrastructure, such as power grids, water supply systems, and transportation networks. These attacks can cause widespread disruption and physical damage. For example, compromising the control systems of a power grid could lead to blackouts affecting millions of people. As autonomous vehicles rise, attackers will seek ways to hijack them. This could lead to accidents, transportation network disruptions, or using vehicles as weapons. Ensuring the security of autonomous systems will be crucial as they become more integrated into everyday life.

Here are some countermeasures:

- **Critical infrastructure protection**: Implement specialized frameworks to secure systems like power grids and water supplies.

- **Autonomous vehicle security**: Develop robust security measures for autonomous vehicles to prevent hijacking and ensure safe operations.

- **Regular security audits**: Conduct regular security audits of critical infrastructure to identify and mitigate vulnerabilities.

Augmented reality and virtual reality exploits

As **Augmented Reality (AR)** and **Virtual Reality (VR)** technologies become integrated into daily life and business operations, they will become new targets for data theft and espionage. Cybercriminals could exploit vulnerabilities in AR/VR platforms to access sensitive information or disrupt virtual experiences. Attackers could create immersive phishing environments within AR/VR platforms, tricking users into revealing sensitive information in a realistic virtual setting. For example, a user in a virtual meeting might be deceived into sharing confidential data with a fake avatar.

Here are some countermeasures:

- **Immersive environment monitoring**: Deploy tools to monitor AR and VR spaces for suspicious activity.

- **User education**: Teach users about the unique risks in AR and VR environments to enhance their awareness and caution.

- **Secure development practices**: Ensure that AR and VR applications are developed with security as a core component to prevent exploits.

5G-enabled attacks

The increased bandwidth and connectivity offered by 5G networks will enable more powerful and disruptive attacks. These attacks can overwhelm networks and services, causing significant downtime and financial losses. Additionally, 5G networks allow for the creation of network slices, which attackers can individually target and exploit. Each slice can be dedicated to a specific service or application; compromising one can have significant implications for the associated service.

Here are some countermeasures:

- **Enhanced DDoS protection**: Implement advanced DDoS protection solutions to defend against high-bandwidth attacks enabled by 5G.

- **Network slice security**: Ensure each network slice is individually secured and monitored to prevent exploitation.

- **Regular network assessments**: Conduct security assessments of 5G networks to identify and address potential vulnerabilities.

Cyber-attacks' future might sound like a techno-thriller nightmare but don't panic. As threats evolve, so do our defenses. The key is staying informed, adaptable, and, yes, a little bit paranoid. A healthy dose of skepticism might be your best firewall in this ever-changing digital landscape. Remember, today's science fiction could be tomorrow's headline in cybersecurity. Stay sharp, stay updated, and think twice before asking your smart toaster to handle your online banking. By implementing these countermeasures, advanced cybersecurity professionals can better prepare for the evolving threat landscape and protect their organizations against sophisticated future threats.

The role of AI and ML in cyber security

The incorporation of AI and ML stands as a beacon of hope for cybersecurity. These technologies have the potential to revolutionize how we defend against cyberattacks, offering enhanced threat detection, rapid response, and a deeper understanding of the dynamic nature of modern cybersecurity challenges. This section will take a comprehensive look at the impactful role of AI and ML in strengthening our cyber defenses.

Navigating the AI landscape – risks, threats, and best practices for secure adoption

AI has become a cornerstone of modern enterprise, offering unprecedented opportunities for innovation and efficiency. However, with these benefits come significant risks and threats that businesses must navigate carefully. This section explores real-world AI risks and scenarios, predicts future AI-related threats, and outlines best practices for secure AI adoption.

The rapid integration of AI into business operations has introduced many new risks. Enterprises must be vigilant about these potential threats. Data protection and privacy are significant concerns, as generative AI tools can inadvertently expose sensitive data. This risk is exacerbated when AI applications are not configured to handle private information securely. Misconfigurations in access controls can lead to unauthorized access to sensitive data, posing risks when AI chatbots generate responses based on historical data from high-level executives. AI-driven cyber threats, such as sophisticated phishing and social engineering attacks, are becoming more prevalent. AI tools can generate convincing phishing emails and deepfakes, making traditional detection methods less effective. Additionally, AI can create polymorphic malware that evolves to evade detection, making it a formidable tool for cybercriminals. Regulatory and compliance challenges also arise as AI technologies advance faster than the development of corresponding regulatory frameworks, especially in industries with stringent data protection requirements.

The nature and scope of AI-driven threats will also change as AI continues to evolve. State-sponsored AI threats are expected to increase, with nation-states leveraging AI to enhance the sophistication and scale of their cyber-attacks. These attacks may target critical infrastructure, intellectual property, and national security interests. The dark web will see a surge in the availability of malicious AI tools such as WormGPT and FraudGPT, which are used to conduct social engineering, phishing, and other cyber-attacks with greater efficacy. Data poisoning, where malicious actors compromise the integrity and reliability of AI models by poisoning training data, will become more common, particularly in critical sectors such as healthcare and finance. AI will also be increasingly used to generate deepfakes and other forms of disinformation, posing significant risks to democratic processes and public trust, including election interference by both domestic and foreign actors.

To harness the benefits of AI while mitigating its risks, enterprises must adopt robust security practices. Conducting comprehensive risk assessments to identify and understand specific threats associated with AI tools is essential. This includes evaluating the potential for data leakage, access control weaknesses, and other vulnerabilities. Robust access control mechanisms, such as **role-based access control** (RBAC) and MFA, ensure that only authorized users can access sensitive AI tools and data. Deploying **Data Loss Prevention** (DLP) solutions to monitor and protect sensitive information is crucial. This includes preventing unauthorized data transfers and ensuring that AI applications do not expose critical data. Regular audits and continuous monitoring of AI systems for signs of misuse or anomalies can help identify potential security gaps and ensure compliance with relevant regulations.

Following secure development practices when creating and deploying AI applications, including performing code reviews, conducting security testing, and using secure coding frameworks, is vital. Employee training and awareness programs can help mitigate the risk of social engineering and phishing attacks by educating employees about the risks associated with AI and the importance of following security protocols. Hosting AI applications in secure, private instances rather than relying on public platforms can help maintain control over data and reduce the risk of exposure. Finally, staying informed about emerging regulations and working closely with regulatory bodies to ensure compliance can help mitigate legal and compliance risks.

Integrating AI into business operations is a double-edged sword, offering transformative potential and significant risks. Enterprises can navigate this complex landscape effectively by understanding real-world threats and adopting best practices for secure AI adoption. As AI technologies evolve, staying ahead of emerging threats and maintaining a robust security posture will be essential for safeguarding data and ensuring the successful deployment of AI solutions.

AI and ML for threat detection

AI and ML's strength in cybersecurity lies in their capacity to process and analyze vast datasets with unparalleled efficiency. Their ability to recognize patterns makes them invaluable in identifying and mitigating threats. Let's delve deeper into how these technologies are making significant strides in threat detection.

Behavior-based anomaly detection

Behavior-based anomaly detection represents a pivotal application of AI and ML in cybersecurity. These technologies, powered by complex algorithms and data analysis, scrutinize network and user behaviors, rapidly detecting deviations that might indicate cyberattacks.

For instance, consider the scrutinizing of login patterns. AI and ML systems, armed with the capability to learn what constitutes *normal* behavior, can swiftly recognize when login patterns deviate from the established norm. Such deviations, especially if numerous failed login attempts are detected, may be indicative of a brute-force attack. This level of nuanced analysis goes beyond the capabilities of traditional signature-based systems, which rely on predefined patterns.

Moreover, AI and ML are proficient at recognizing zero-day threats, a formidable challenge in the cybersecurity landscape. Unlike signature-based systems that depend on known patterns, AI and ML systems learn from historical data and adapt to evolving threats. They can identify deviations in behavior that do not match predefined patterns, making them adept at spotting novel attacks. An example of this prowess can be seen in advanced spearphishing attacks. AI and ML methods excel at identifying anomalous user behaviors, such as irregular email communication, that could signify phishing attempts and other cyber threats.

Advanced malware detection

AI and ML have become the linchpins in advanced malware detection. With the continuous evolution of malware, from traditional viruses to polymorphic and fileless malware, conventional signature-based detection struggles to keep pace. AI and ML step in by analyzing code behavior, file attributes, and network activities to identify previously unknown malware, providing a robust defense against the ever-evolving landscape of threats.

ML models, a subset of AI, are particularly adept at this task. They can analyze the code behavior of files, looking for subtle patterns that might indicate malicious intent. This approach enables the detection of malware even if it has never been encountered before. Moreover, AI and ML excel at discerning suspicious network traffic patterns that may indicate malware communication. For instance, unusual spikes in data transmission to unknown destinations can be flagged as potential threats.

A study by Saxe and Berlin (`https://arxiv.org/abs/1508.03096`) highlighted the potential of deep neural network-based malware detection methods. These methods leverage AI and ML to scrutinize code behavior, showcasing the efficacy of ML in identifying malicious code behavior. This represents a significant step forward in defending against rapidly evolving malware threats.

Zero-day vulnerability detection

One of the most critical challenges in cybersecurity is identifying zero-day vulnerabilities, which are previously unknown weaknesses in software. These vulnerabilities often serve as gateways for cyberattacks, and their discovery is of paramount importance in safeguarding systems and data.

AI and ML excel in recognizing patterns from historical data, a capability that makes them ideal for identifying potential vulnerabilities. By analyzing extensive datasets that include historical attack data and patterns, ML models can pinpoint vulnerabilities in code or system configurations. These identified weaknesses may not have been previously known, but through pattern recognition and data analysis, AI and ML can flag them as potential risks.

Organizations can then focus on proactive patching or implement compensatory measures to mitigate the risk associated with these vulnerabilities. This approach aligns with a study in 2017, which surveyed the role of ML in handling big data and recognizing patterns that indicate vulnerabilities. It underscores the importance of utilizing AI and ML to enhance the security posture of organizations by identifying and addressing vulnerabilities before they are exploited by malicious actors.

The evolving threat landscape

Cybersecurity is a dynamic field, with adversaries constantly adapting and evolving their tactics. Understanding the trajectory of emerging threats is vital for organizations to bolster their defenses effectively:

- **Advanced Persistent Threats (APTs)**: APTs represent a significant shift in cyber threats. These attacks are characterized by their persistence, sophistication, and often state-sponsored backing. Nation-state actors, organized crime groups, and even hacktivists employ APTs to achieve various objectives, from espionage to sabotage. APTs leverage multiple attack vectors and often employ a combination of social engineering, zero-day exploits, and custom malware to infiltrate and maintain long-term access to target systems. Their advanced tactics make them elusive and challenging to detect. Understanding their tactics and motivations is crucial for organizations.

APTs aim to remain undetected for extended periods, quietly exfiltrating sensitive data or maintaining persistent access for future attacks. They target specific organizations or industries based on their strategic interests, making each APT campaign unique.

- **Nation-state actors**: In recent years, nation-states have become increasingly active in the cyber domain. Governments worldwide have recognized the strategic advantages of cyber operations for espionage, influencing international affairs, and strategic advantage. Nation-state actors often target not only government entities but also private organizations, critical infrastructure, and even elections in other countries.

  These state-sponsored groups are well-funded, highly skilled, and often possess extensive resources, which they leverage to conduct cyber-espionage, launch disruptive attacks, or exert influence on foreign policy. Recognizing the involvement of nation-state actors in cyberattacks is paramount for organizations and governments as they respond to and mitigate these threats.

- **Ransomware evolution**: Ransomware attacks have witnessed a significant evolution. What once started as opportunistic attacks, targeting individual users and small businesses, has now evolved into highly targeted campaigns that seek to maximize financial gain. Understanding these changes is critical for organizations as they prepare for the ransomware threat.

  Modern ransomware attacks are characterized by their ability to infiltrate large organizations, encrypt critical data, and demand substantial ransoms for decryption keys. Some attackers even threaten to release stolen data, further pressuring victims to pay. Understanding the motivations and tactics of ransomware groups is key to developing effective defenses and response strategies.

Having explored the role of AI and ML in cybersecurity, let's try to understand the emerging threat vectors and the advanced strategies required to mitigate them.

Emerging threat vectors

The threat landscape is continually evolving, driven by advances in technology and changes in attacker tactics. Organizations must stay ahead of emerging threat vectors to adapt their defenses effectively:

- **IoT vulnerabilities**: The proliferation of IoT devices has introduced a host of new vulnerabilities into the cybersecurity landscape. IoT devices, often designed with a focus on functionality rather than security, frequently lack robust security measures.

  Cybercriminals are exploiting these weaknesses, targeting IoT devices for a range of malicious activities, including launching DDoS attacks, creating botnets, and gaining unauthorized access to networks. Understanding the security challenges posed by IoT devices is essential for organizations to secure their digital ecosystems effectively.

- **AI and ML in cyberattacks**: AI and ML are no longer tools just limited to defenders; they have become integral to attackers' arsenals. Cybercriminals are increasingly employing AI and ML for more sophisticated attacks, including credential stuffing, adaptive phishing campaigns, and evasion of traditional security measures. AI-driven attacks can automate tasks, identify vulnerabilities, and adapt to changing conditions. Understanding the role of AI and ML in cyber threats is vital for organizations to develop defenses that can counter these advanced techniques effectively.

- **Supply chain attacks**: Attackers have increasingly recognized the potential of targeting the supply chain to compromise organizations indirectly. In supply chain attacks, adversaries infiltrate trusted third-party vendors or service providers and use their access to compromise their actual target. Supply chain attacks can lead to the distribution of malicious software or the theft of sensitive data from the primary target.

  Recognizing the risks associated with supply chain attacks is crucial for organizations to secure their supply chain and ensure the integrity and security of the products and services they receive.

Mitigation and preparedness strategies

As the threat landscape evolves, organizations must adapt their cybersecurity strategies and preparedness to stay resilient against emerging threats. This entails a proactive approach to threat mitigation and defense:

- **Threat intelligence**: Leveraging threat intelligence feeds and services can help organizations stay informed about emerging threats. Threat intelligence provides real-time information about emerging cyber threats, tactics, and vulnerabilities. Organizations can use this data to adjust their defenses and make informed decisions about their security posture.

  Threat intelligence sources can include government agencies, industry-specific **information sharing and analysis centers** (**ISACs**), and private cybersecurity companies. Collaborating with these entities and sharing threat information can enhance collective defense.

- **Awareness and training**: Cybersecurity education and training are fundamental components of an effective defense strategy. Users and employees are often the first line of defense against cyber threats. They need to be aware of the latest threats and best practices to protect themselves and their organizations.

  Security awareness programs simulated phishing exercises, and regular training sessions can help educate users about emerging threats, including phishing, social engineering, and ransomware. The goal is to empower users to recognize and report suspicious activities and practice safe online behaviors.

- **Incident response planning**: Incident response is a critical aspect of cybersecurity preparedness. Organizations should develop robust incident response plans that consider emerging threat scenarios. These plans should outline clear procedures for detecting, responding to, and recovering from cyber incidents.

  Effective incident response includes establishing an incident response team, defining roles and responsibilities, developing communication and notification procedures, and regularly testing the plan through tabletop exercises and simulations.

Predicting the future of cyber threats is a challenging but necessary endeavor. As APTs, nation-state actors, and emerging threat vectors continue to evolve, organizations must remain vigilant, adapt their defenses, and prioritize cybersecurity to stay ahead of the rapidly changing threat landscape.

By understanding the intricacies of these emerging threats and implementing effective mitigation and preparedness strategies, organizations can enhance their security posture and remain resilient in the face of the ever-evolving cybersecurity landscape. Vigilance, collaboration, and continuous learning are key to staying one step ahead of cyber adversaries.

Future trends in cybersecurity

As cybersecurity continues to evolve, several trends are expected to shape the future of the field:

AI and ML are anticipated to play an increasingly significant role in cybersecurity. These technologies have the ability to analyze vast datasets and identify patterns indicative of cyber threats. ML can enhance the accuracy and speed of threat detection and response.

Quantum computing, while heralded for its potential to revolutionize various fields, also poses a significant threat to cybersecurity. Quantum computers could theoretically break current encryption standards. As a result, quantum-resistant cryptography is emerging as a critical field to ensure the security of sensitive data in a post-quantum world.

The evolution of cybersecurity models and strategies is an intricate journey that reflects the dynamic nature of the cybersecurity landscape. With every new threat, the realm of cybersecurity adapts and evolves to counter the challenges. As we move forward, organizations and cybersecurity professionals must stay informed about emerging trends and be prepared to adapt to new and sophisticated threats. The evolution of cybersecurity is a testament to the resilience and ingenuity of the field as it continues to defend against the ever-evolving landscape of cyber threats.

This extended content now provides a detailed examination of the evolution of cybersecurity models, the challenges that prompted these changes, and the strategies that have emerged to address new threats. It also highlights future trends that are expected to shape the field of cybersecurity. Please let me know if you need any further information or if you have specific requirements for this section.

Automation and threat response

The speed and scale of modern cyber threats demand automated response systems that can act swiftly and decisively:

- **Automated incident response**: AI and automation play a significant role in automating incident response, enabling rapid reactions to security incidents. These systems can isolate affected systems, block malicious activities, and initiate predefined incident response plans. AI-driven automation helps reduce the time it takes to detect and respond to incidents, limiting the impact and reducing the overall risk.

- **Threat intelligence analysis**: Continuous monitoring and analysis of threat intelligence feeds are paramount for staying proactive against emerging threats. AI systems can sift through an abundance of threat data, identifying patterns, and indicators of compromise, and assessing the relevance of the threats to the organization. The integration of AI and ML in this domain significantly reduces the human effort required for threat analysis. Organizations can leverage automated tools to ensure that the threat landscape is consistently monitored and acted upon.

As the future of cybersecurity rapidly evolves with advancements in AI-driven technologies, it brings forth unprecedented opportunities for innovation and protection. However, with these advancements come pressing ethical and legal challenges that must be carefully considered in the intersection of AI and cybersecurity.

Ethical and legal implications of AI and cybersecurity

As AI and ML continue to revolutionize cybersecurity, they bring a host of ethical and legal considerations that advanced cybersecurity professionals must navigate. The deployment of these transformative technologies raises questions about privacy, fairness, accountability, and regulatory compliance and inspires us to rethink and reshape the cybersecurity landscape. This section delves into these implications, exploring how they impact the cybersecurity landscape and what professionals must consider as they integrate AI and ML into their security frameworks:

- **Privacy concerns**: AI and ML technologies often require vast amounts of data to function effectively. This data typically includes sensitive personal information, which raises significant privacy concerns. The use of AI in cybersecurity must balance the need for comprehensive data analysis with the imperative to protect individual privacy. Organizations must ensure data collection practices comply with privacy regulations such as the **General Data Protection Regulation** (**GDPR**) and the **California Consumer Privacy Act** (**CCPA**). Implementing robust data anonymization and encryption techniques can help mitigate privacy risks while enabling effective AI-driven security measures.

- **Algorithmic bias**: One of the critical ethical challenges in AI is the potential for algorithmic bias. AI systems are only as unbiased as the data on which they are trained. If the training data contains biases, the AI system can perpetuate and even amplify these biases, leading to unfair or discriminatory outcomes. In cybersecurity, certain groups or behaviors are disproportionately flagged as suspicious based on biased data. Advanced cybersecurity professionals must not only rigorously test and validate their AI models but also continuously monitor their performance to ensure they are fair and unbiased. This proactive approach involves using diverse datasets and being transparent about the AI's decision-making processes.

- **Accountability and transparency**: The opacity of AI decision-making processes, often called the **black box** problem, presents challenges for accountability and transparency. In cybersecurity, where decisions can have significant implications, it is crucial to understand how and why an AI system makes certain decisions. Ensuring transparency involves the following:

  - Documenting AI algorithms

  - Maintaining logs of AI decision-making processes

  - Being able to explain these decisions to stakeholders

  Accountability measures should also include clear protocols for addressing and rectifying errors or biases detected in AI systems.

- **Regulatory compliance**: The regulatory landscape for AI in cybersecurity is rapidly evolving. Governments and regulatory bodies are increasingly focusing on the implications of AI technologies, leading to the development of new laws and guidelines. Advanced cybersecurity professionals must stay informed about these regulatory changes to ensure compliance. This includes understanding regulations related to data protection, AI ethics, and cybersecurity standards. More importantly, engaging with regulatory bodies and participating in industry discussions can help organizations not only stay ahead of regulatory requirements but also influence future policy developments, making them an integral part of the regulatory process.

- **Ethical use of AI in cyber defense**: While AI offers powerful tools for enhancing cybersecurity, ethical principles must guide its use. This includes respecting user privacy, ensuring fairness, and avoiding misuse of AI capabilities. For instance, using AI for surveillance or monitoring should be done transparently and with the consent of the individuals being monitored. Ethical guidelines should be established and adhered to within organizations, ensuring that the deployment of AI in cybersecurity aligns with broader societal values and moral standards.

- **Balancing security and privacy**: One of the enduring challenges in cybersecurity is balancing the need for security with the protection of individual privacy rights. AI systems amplify this challenge with their ability to process and analyze vast amounts of data. Advanced cybersecurity professionals must develop strategies that achieve this balance, such as implementing privacy-by-design principles, where privacy considerations are integrated into every stage of AI development and deployment.

- **Future ethical challenges**: Looking ahead, the ethical landscape of AI in cybersecurity will become even more complex. Issues such as AI autonomy, where AI systems operate with minimal human intervention, raise questions about control and oversight. The potential for AI to be used in offensive cyber operations also poses significant ethical dilemmas. Advanced cybersecurity professionals must proactively engage with these emerging ethical challenges, contributing to developing ethical frameworks and best practices that guide the responsible use of AI in cybersecurity.

Integrating AI and ML into cybersecurity brings transformative potential and significant ethical and legal implications. Advanced cybersecurity professionals must navigate these challenges by ensuring privacy, mitigating algorithmic bias, maintaining accountability and transparency, and staying compliant with evolving regulations. By adhering to ethical principles and contributing to the development of responsible AI practices, professionals can harness the power of AI to enhance cybersecurity while upholding the values of fairness, privacy, and accountability.

The human factor in cybersecurity

In the rapidly evolving landscape of cybersecurity, technology often takes center stage. However, the human element remains a crucial factor in the effectiveness of security measures. As we look to the future, the role of human behavior and psychology in cybersecurity will become even more pronounced. Understanding and mitigating human vulnerabilities will strengthen defenses against increasingly sophisticated cyber threats. This section explores the evolving role of human factors in cybersecurity, emphasizing the importance of human-centric strategies in combating cyber threats:

- **Human vulnerabilities in cybersecurity**: Despite technological advances, human error continues to be one of the most significant vulnerabilities in cybersecurity. Phishing attacks, for example, exploit human trust and curiosity. In 2020, a significant data breach at Twitter occurred when attackers used social engineering techniques to trick employees into revealing their credentials. This incident underscores the critical role human behavior plays in the security landscape. The following are the two common types of human vulnerabilities:

  - **Cognitive biases**: Human cognitive biases such as overconfidence can lead to risky behaviors, including reusing passwords or ignoring security warnings. Cybercriminals often exploit these biases through social engineering tactics.

  - **Insider threats**: Malicious insiders or negligent employees can cause significant damage. The 2017 Equifax breach, which exposed the personal data of 147 million people, was partially attributed to a failure to apply a critical security update to an employee.

- **Emerging trends in social engineering attacks**: Social engineering attacks are evolving, leveraging technological advancements and deepening insights into human psychology. Attackers use sophisticated techniques to manipulate individuals into divulging sensitive information or performing actions compromising security. The following trends are being observed nowadays in terms of social engineering attacks:

  - **Deepfake technology**: Deepfake technology, which uses AI to create realistic but fake audio and video content, is used in phishing and spear-phishing attacks. In one notable case, cybercriminals used a deepfake voice to impersonate a CEO and trick an employee into transferring $243,000 to a fraudulent account.

- **Personalized attacks:** With the vast amount of personal information available online, attackers can craft highly personalized and convincing phishing emails. By analyzing social media activity, purchasing habits, and professional networks, cybercriminals create messages that are difficult to distinguish from legitimate communications.

- **Thought-provoking questions:** To inspire critical thinking and prompt discussions on the evolving landscape of cybersecurity about human factors, consider the following questions:

  - How can organizations balance the need for security with the user experience to ensure compliance without creating frustration?

  - What role does leadership play in fostering a security culture within an organization, and how can leaders effectively drive this culture change?

  - How can advanced cybersecurity professionals stay ahead of emerging social engineering tactics and continuously improve their defenses?

The human factor in cybersecurity is a dynamic and critical component of an organization's overall security posture. As cyber threats evolve, so must our understanding and mitigation of human vulnerabilities. Organizations can significantly enhance their defenses by implementing comprehensive user awareness programs, leveraging behavioral analytics, and fostering a robust security culture. Advanced cybersecurity professionals must remain vigilant, adaptable, and proactive in addressing the human element to combat the sophisticated threats of the future effectively.

Blockchain and cybersecurity

Blockchain technology, with its foundational principles of decentralization, immutability, and transparency, is on the brink of revolutionizing the field of cybersecurity. Originally designed to support cryptocurrencies such as Bitcoin, blockchain's unique attributes offer innovative solutions to many contemporary security challenges. This section delves into the advanced applications of blockchain in cybersecurity, providing insights for professionals on how to leverage this technology to enhance security protocols and architectures, sparking excitement and optimism about the future of cybersecurity:

- **Decentralized identity management:** Decentralized identity management systems based on blockchain provide a robust alternative to traditional centralized systems, often vulnerable to single points of failure and data breaches. Blockchain enables users to control their identities, reducing the risk of identity theft and fraud. For instance, Microsoft's Azure Active Directory decentralized identity solutions leverage blockchain to offer users greater control over their data, enhancing privacy and security. This shift toward user-centric identity management is critical for mitigating risks associated with centralized identity repositories.

- **Immutable audit trails:** The immutability of blockchain records ensures that once data is recorded, it cannot be altered or tampered with, providing a reliable foundation for creating tamper-proof logs and audit trails. This reliability is essential for regulatory compliance and forensic investigations. For example, Guardtime's blockchain-based system protects national infrastructure by ensuring the integrity and provenance of data, creating an immutable log of digital transactions. This system, used by the Estonian government (`https://e-estonia.com/solutions/cyber-security/ksi-blockchain/`), showcases how blockchain can enhance the transparency and reliability of digital records, providing a sense of security and reassurance to professionals in the field.

- **Smart contracts and automated security policies**: Smart contracts—self-executing contracts with the terms of the agreement directly written into code—can automate and enforce security policies without human intervention. This reduces the likelihood of human error and ensures consistent policy enforcement. For instance, smart contracts can automate the execution of security protocols, such as access control policies and incident response procedures, ensuring that these actions are carried out precisely and without delay.

- **Real-world applications**: A notable example of blockchain's application in cybersecurity is the partnership between IBM and Maersk to create TradeLens, a blockchain-based shipping platform. TradeLens enhances security by providing a transparent and immutable record of shipping transactions, reducing the risk of fraud and ensuring the integrity of shipping data.

As blockchain continues to revolutionize cybersecurity with decentralized security models, the growing adoption of cloud technologies presents its own unique set of challenges and opportunities in safeguarding digital asset

Cybersecurity and the cloud

Cloud computing has fundamentally reshaped the IT landscape, offering unparalleled scalability, flexibility, and cost efficiency. However, as organizations increasingly migrate their critical workloads and sensitive data to the cloud, the cybersecurity challenges become more complex and nuanced. This section explores advanced concepts and cutting-edge developments in cloud security, providing insights for advanced cybersecurity professionals on how to stay ahead in this rapidly evolving field. Here are some key technologies to consider:

- **Advanced encryption techniques**: Encryption remains a cornerstone of cloud security, but traditional methods are being augmented by more sophisticated techniques, such as the following, to address emerging threats:

  - **Homomorphic encryption**: This technique allows computations on encrypted data without decrypting it, thereby maintaining data confidentiality throughout the processing life cycle. This is particularly beneficial for secure data processing and analytics in the cloud. For instance, Google Cloud is actively researching and developing homomorphic encryption solutions to enhance data security in its services.

- **Quantum-resistant cryptography**: With the advent of quantum computing, traditional encryption methods are at risk. Quantum-resistant cryptographic algorithms are being developed to safeguard data against the immense processing power of future quantum computers. Organizations such as IBM and NIST are leading efforts to standardize these new cryptographic methods.

- **AI-driven security solutions**: AI and ML are revolutionizing cloud security by enabling real-time threat detection and automated response mechanisms in the following ways:

  - **Anomaly detection**: AI-driven systems can analyze vast amounts of data to establish baseline behaviors and identify anomalies that may indicate security threats. AWS GuardDuty, for example, uses ML to monitor network traffic, user behavior, and other metrics to detect unusual patterns and potential security incidents.

  - **Automated incident response**: Leveraging AI for incident response can significantly reduce reaction times and mitigate the impact of security breaches. Computerized systems can trigger predefined responses, such as isolating compromised instances or blocking suspicious IP addresses, ensuring swift containment of threats.

- **Cloud Access Security Brokers (CASBs)**: CASBs have become essential tools for enforcing security policies and ensuring compliance in cloud environments.

  - **Visibility and control**: CASBs provide comprehensive visibility into cloud usage across an organization, enabling security teams to monitor and manage access to cloud resources. Netskope and Microsoft Cloud App Security offer detailed insights into user activities, data transfers, and application usage.

  - **DLP**: By integrating with CASBs, organizations can enforce DLP policies to protect sensitive data from unauthorized access and exfiltration. These tools can automatically encrypt or block the transfer of sensitive information based on predefined rules, ensuring compliance with regulatory requirements.

- **Integration with DevSecOps**: Integrating security into the DevOps process, known as DevSecOps, is critical for ensuring that security is a fundamental part of the cloud application life cycle:

  - **Automated security testing**: Integrating automated security testing into the **continuous integration/deployment (CI/CD)** pipeline ensures that vulnerabilities are identified and addressed early in development. Tools such as Snyk and Veracode provide automated scanning for code vulnerabilities, enabling developers to fix issues before applications are deployed.

  - **Infrastructure as Code (IaC) security**: Organizations can apply the same security controls and testing mechanisms in application development by treating infrastructure configurations as code. IaC tools such as Terraform and AWS CloudFormation allow for automated and consistent deployment of secure infrastructure configurations.

- **Future trends in cloud security**: As cloud technology continues to evolve, a couple of trends are expected to shape the future of cloud security:

  - **Serverless security**: With the rise of serverless architectures, new security challenges emerge, such as securing ephemeral functions and managing the increased attack surface. Advanced security solutions are being developed to provide visibility and protection for serverless environments.

  - **Edge computing security**: As edge computing becomes more prevalent, securing data and applications at the edge will be crucial. This involves protecting distributed networks of edge devices and ensuring secure data transmission between edge locations and central cloud services.

The future of cloud security is marked by continuous innovation and adaptation to emerging threats. Advanced cybersecurity professionals must stay informed about the latest developments in encryption, AI-driven security, CASBs, and DevSecOps to effectively protect cloud environments. By leveraging cutting-edge technologies and best practices, professionals can ensure robust security and compliance in an increasingly complex and dynamic cloud landscape.

Cybersecurity skills and workforce development

As the cybersecurity landscape evolves, the demand for skilled professionals outpaces supply. Emerging technologies such as blockchain, critical infrastructure protection, and cloud security require specialized knowledge and expertise. This section explores the advanced concepts and cutting-edge developments in cybersecurity skills and workforce development, emphasizing the need for continuous learning and adaptation to stay ahead of emerging threats.

Building a skilled cybersecurity workforce

Addressing the cybersecurity skills gap requires a multifaceted approach, including formal education, certifications, hands-on experience, and continuous professional development:

- **Formal education and certifications**: Advanced degree programs in cybersecurity, computer science, and related fields provide foundational knowledge and specialized training. Certifications such as *CISSP*, *CISM*, and *CEH*, and cloud-specific credentials such as *AWS Certified Security—Specialty* and *Microsoft Certified: Azure Security Engineer Associate* validate expertise and enhance credibility.

- **Hands-on experience**: Practical experience is invaluable in developing advanced cybersecurity skills. Simulation environments, such as cyber ranges and **Capture the Flag** (**CTF**) competitions, provide real-world scenarios for professionals to hone their skills. Internships, co-op programs, and job rotations within cybersecurity teams also offer critical hands-on experience.

- **Continuous professional development**: The dynamic nature of cybersecurity necessitates ongoing learning. Attending conferences, participating in webinars, enrolling in online courses, and engaging with professional communities help professionals stay current with the latest trends, tools, and techniques.

Strategies for workforce development

Organizations must implement strategic initiatives to attract, retain, and develop cybersecurity talent:

- **Talent pipeline development**: Collaborating with educational institutions to create tailored curricula that align with industry needs can help build a robust talent pipeline. Offering scholarships, internships, and mentorship programs can attract new talent.

- **Upskilling and reskilling**: Providing opportunities for current employees to upskill and reskill is essential. Internal training programs, cross-functional projects, and access to advanced training resources enable employees to expand their skill sets and adapt to new organizational roles.

- **Diversity and inclusion**: Promoting diversity and inclusion within the cybersecurity workforce enhances problem-solving and innovation. Organizations should strive to create an inclusive environment that supports diverse perspectives and experiences, which are critical in addressing complex cybersecurity challenges.

The role of emerging technologies in workforce development

The following emerging technologies are shaping cybersecurity practices and influencing workforce development strategies:

- **AI and ML**: AI and ML transform cybersecurity by automating threat detection, response, and predictive analysis. Professionals need to develop skills in AI and ML to design, implement, and manage these technologies within security operations.

- **Automation and orchestration**: The increasing use of automation and orchestration tools requires professionals to be proficient in scripting, workflow management, and integrating automated processes into security operations. Understanding how to leverage tools such as **Security Orchestration, Automation, and Response (SOAR)** platforms is crucial.

- **Quantum computing**: As quantum computing advances, cybersecurity professionals must prepare for its impact on cryptography. Developing quantum-resistant algorithms and understanding quantum cryptographic principles will be essential skills shortly.

The future of cybersecurity relies heavily on a skilled and adaptable workforce capable of navigating the complexities of emerging technologies and evolving threats. Advanced cybersecurity professionals must continuously expand their knowledge and expertise through formal education, certifications, hands-on experience, and ongoing professional development. By fostering a culture of continuous learning and investing in workforce development strategies, organizations can build resilient and innovative cybersecurity.

Continuous improvement in cyber defense strategies

The necessity of continuous improvement in cyber defense strategies cannot be overstated in the face of the rapidly evolving cyber threat landscape. It is imperative for cybersecurity professionals to constantly update their knowledge and integrate the latest advancements into their defense mechanisms. This section explores advanced concepts and cutting-edge developments that are shaping the future of cybersecurity, providing a comprehensive guide for seasoned experts to enhance their defense strategies.

Adaptive threat intelligence

Adaptive threat intelligence represents a significant leap in proactive cybersecurity measures. Unlike traditional threat intelligence, which relies on historical data, adaptive threat intelligence leverages ML and AI to predict and respond to threats in real time. By continuously analyzing vast amounts of data from various sources, these systems can identify patterns and anomalies indicative of potential threats. For instance, AI-driven platforms such as Darktrace use unsupervised learning to detect and respond to novel threats autonomously, reducing the time to remediation and improving overall security posture.

Zero Trust architecture

Zero Trust Architecture (**ZTA**) has emerged as a critical framework for modern cyber defense. Rooted in the principle of "never trust, always verify," ZTA eliminates implicit trust within the network, requiring continuous verification of every device, user, and application. Implementing ZTA involves micro-segmentation, robust **identity and access management** (**IAM**), and constant monitoring. For example, Google's BeyondCorp initiative exemplifies ZTA by allowing employees to work securely from any location without needing a traditional VPN, ensuring stringent security controls are applied consistently across the network.

Advanced encryption techniques

Encryption remains a cornerstone of cybersecurity, but advancements in quantum computing necessitate the development of quantum-resistant algorithms. **Post-quantum cryptography** (**PQC**) aims to create encryption methods that can withstand quantum attacks. NIST is leading the charge in standardizing PQC algorithms, with promising candidates such as lattice-based, hash-based, and multivariate polynomial cryptography. As quantum computers edge closer to reality, adopting PQC will be critical in safeguarding sensitive data.

Behavioral analytics and insider threat detection

Insider threats pose a significant challenge to cybersecurity, often bypassing traditional defenses. Behavioral analytics offers a sophisticated approach to detecting insider threats by monitoring user behavior and identifying deviations from established norms. Solutions such as Splunk's **User Behavior Analytics** (**UBA**) leverage ML to analyze activities across multiple dimensions, such as login patterns, data access, and communication behaviors. By flagging anomalous actions, these tools enable organizations to address potential insider threats before they escalate preemptively.

Automation and orchestration

The complexity and volume of cyber threats necessitate automation and orchestration to streamline incident response. SOAR platforms integrate various security tools and automate repetitive tasks, allowing security teams to focus on higher-order analysis. Platforms such as Palo Alto Networks' Cortex XSOAR provide comprehensive automation capabilities, from threat hunting to incident response, enabling faster and more effective mitigation of threats. Automation not only enhances efficiency but also ensures consistency in response actions.

Deception technology

Deception technology is an innovative approach that uses decoys and traps to lure attackers away from critical assets. Organizations can detect, analyze, and respond to malicious activities without exposing real systems by creating a network of false targets. Solutions such as Attivo Networks' ThreatDefend platform employ dynamic deception tactics, deploying decoys that mimic real assets to engage and study attackers. This proactive strategy confuses and slows down adversaries and provides valuable intelligence on their tactics and techniques.

AI in cyber defense

While integrating AI and ML marks a significant leap in enhancing cybersecurity, it's crucial to recognize that the battle against cyber threats is ongoing. The threat landscape continually evolves, necessitating adaptable defense strategies. AI continues to revolutionize cybersecurity, offering capabilities that extend beyond human limitations. AI-driven tools excel in threat detection, predictive analytics, and automated response. Deep learning models, such as those employed by Cylance, analyze vast datasets to identify previously unknown malware and sophisticated attack vectors. Furthermore, AI enhances threat hunting by correlating data across disparate sources, uncovering hidden threats that traditional methods might miss. Integrating AI in cybersecurity operations promises to elevate defense mechanisms to unprecedented efficacy.

Adaptability and evolution in cybersecurity

This section explores the latest advancements in AI and ML technologies and their impact on cybersecurity, highlighting both the opportunities and challenges they present.

The dynamic cyber threat landscape

The first and most crucial aspect of continuous improvement in cybersecurity is maintaining a deep understanding of the dynamic threat landscape. Cyber threats are not static; they evolve continuously, becoming more sophisticated and elusive. The technologies that protect against them must also evolve to keep pace. This dynamic nature of the cyber threat landscape underscores the need for constant vigilance and adaptability in cybersecurity strategies.

To effectively defend against evolving threats, organizations need to stay informed about emerging threats and vulnerabilities. Threat intelligence sources, such as ISACs and open collaboration among security professionals, play a pivotal role in staying ahead of the threat curve. Verizon's *2020 Data Breach Investigations Report* states that awareness of the types of threats prevalent and the tactics they employ is fundamental to crafting an effective defense.

Incident post-mortems

Security incidents, whether successful or thwarted, provide invaluable lessons. Conducting post-incident analyses can reveal vulnerabilities, misconfigurations, and exploited attack vectors. These findings are essential for identifying weaknesses and areas for improvement in cybersecurity strategies.

Post-incident analyses thoroughly examine the attack life cycle, from the initial intrusion to the final exfiltration of data. The goal is to reconstruct the attack chain and identify where security measures failed or could be strengthened. This process often involves forensic examination, including system logs, analyzing malware artifacts, and interviewing personnel involved in incident response.

Proactive threat hunting

Proactive threat hunting, a vital element of continuous improvement in cybersecurity, is highlighted here. While automated systems are crucial for baseline security, they cannot catch everything. Threat hunters and skilled security professionals actively search for hidden threats and vulnerabilities that may evade automated detection. This proactive approach underscores the importance of being one step ahead of cybercriminals.

Threat hunting involves proactively seeking anomalies and traces of malicious activity within the network. It goes beyond automated alerts and focuses on identifying threats lurking undetected. Threat hunters leverage their expertise and tools to actively seek out and neutralize potential threats before they are exploited. Threat hunting is an ongoing and dynamic process. It involves continuously evolving **tactics, techniques, and procedures** (**TTPs**) as cyber adversaries adapt to security measures. Organizations can stay one step ahead of cybercriminals by proactively identifying emerging threats and novel attack techniques.

Cybersecurity mesh architecture

The **cybersecurity mesh architecture** (**CSMA**) is a distributed architectural approach that enables scalable, flexible, and reliable cybersecurity control. CSMA integrates disparate security services into a cohesive ecosystem, ensuring comprehensive visibility and control across various environments, including cloud, on-premises, and hybrid. This approach facilitates a unified policy management system and real-time threat detection, enhancing the organization's ability to respond to evolving threats dynamically. Gartner predicts that by 2025, CSMA will support over half of digital access control requests, underscoring its growing importance in modern cybersecurity strategies.

The future of cybersecurity hinges on continuous improvement and adaptation to emerging threats. By embracing advanced concepts such as adaptive threat intelligence, ZTA, PQC, behavioral analytics, automation, deception technology, AI, and CSMA, organizations can fortify their defenses and stay ahead of adversaries. For cybersecurity professionals, staying informed and agile in adopting these cutting-edge developments is crucial to maintaining robust security in an ever-changing threat landscape. As cyber threats evolve, so must our strategies and technologies to ensure a resilient and secure digital future.

Integrating AI and ML has ushered in a new era in cybersecurity, offering more efficient threat detection and rapid incident response. Nevertheless, the ever-changing threat landscape requires organizations to adapt and continuously improve their cybersecurity strategies. By staying informed about emerging threats, conducting thorough post-incident analyses, and actively hunting for hidden threats, organizations can bolster their defenses against the dynamic and evolving world of cyber threats. AI and ML are pivotal tools in the cybersecurity arsenal. Still, they are most effective when paired with adaptable strategies and a continuous commitment to staying one step ahead of those who seek to exploit vulnerabilities. This combination offers a robust defense against the ever-changing landscape of cyber threats, ensuring that organizations remain resilient, and their security teams are ready to tackle future challenges.

Summary

As the cybersecurity landscape evolves, continuous improvement in defense strategies is critical to counter modern cyber threats' sophisticated and dynamic nature. Integrating AI and ML marks a significant advancement in threat detection and response. AI, with its ability to learn from data, can be used for behavior-based anomaly detection, while ML can aid in advanced malware identification and zero-day vulnerability detection. However, the battle against cyber threats is ongoing, requiring adaptable defense strategies to stay ahead. Understanding the dynamic threat landscape through sources such as threat intelligence and conducting post-incident analyses are fundamental for identifying vulnerabilities and improving cybersecurity measures. Proactive threat hunting by skilled security professionals further enhances an organization's ability to detect and neutralize hidden threats.

Emerging technologies and methodologies are reshaping the cybersecurity paradigm. ZTA, which operates on the principle of "*never trust, always verify*," is essential in modern network security. Advanced encryption techniques, including PQC, are necessary to safeguard data against the looming threat of quantum computing. Deception technology, decoys and traps, and AI-driven automation in incident response are pivotal in modern defense strategies. CSMA integrates disparate security services, providing comprehensive visibility and control across various environments and enhancing real-time threat detection and response.

The future of cybersecurity demands a combination of technological prowess and an adaptive mindset. Organizations must leverage advanced concepts such as adaptive threat intelligence, behavioral analytics, and automation to fortify their defenses. The continuous improvement of cybersecurity strategies involves the following:

- Staying informed about emerging threats

- Conducting thorough post-incident analyses

- Actively hunting for hidden threats

By embracing these advancements, cybersecurity professionals can ensure a resilient and secure digital future, ready to tackle the ever-evolving landscape of cyber threats. AI and ML, while pivotal, are most effective when paired with adaptable strategies and a commitment to staying one step ahead of adversaries. In the next chapter, we will talk about the importance of a proactive approach in the world of cyber security.

Further reading

Following are some bonus reading materials for you to check out:

- *Blog post arguing fingerprints are usernames, not passwords*: `https://blog.dustinkirkland.com/2013/10/fingerprints-are-user-names-not.html`

- *Article explaining adversarial attacks on AI and machine learning*: `https://www.paloaltonetworks.com/cyberpedia/what-are-adversarial-attacks-on-AI-Machine-Learning`

- *Research paper on malware detection using deep neural networks*: `https://arxiv.org/abs/1508.03096`

- *Academic paper on threat modeling for less cyber-dependent adversaries*: `https://papers.academic-conferences.org/index.php/eccws/article/view/2462`

12

A Proactive Approach

Advocating a proactive approach to cybersecurity is important. In today's rapidly evolving digital landscape, the specter of cyberattacks looms more significant than ever before. As technology advances at an unprecedented pace, so too do the tactics and sophistication of cybercriminals. Once considered adequate, traditional reactive measures now falter in the face of increasingly complex and persistent threats. To effectively safeguard our digital assets, a paradigm shift toward proactive cybersecurity is not just beneficial – it's imperative. This proactive approach not only enhances security but also fosters a culture of continuous learning and innovation, positioning organizations to thrive in the digital age.

This chapter explores the critical importance of fostering a proactive cybersecurity mindset. We will delve into the core principles of this approach, emphasizing the need for continuous learning, the anticipation of emerging threats, and the cultivation of a security-aware culture that permeates every level of an organization.

Proactive cybersecurity transcends the mere deployment of cutting-edge technologies or the establishment of robust defenses. At its heart, it's about staying several steps ahead of potential threats, anticipating the next moves of cybercriminals, and being thoroughly prepared to respond swiftly and effectively when incidents inevitably occur. This approach demands a comprehensive understanding of the current threat landscape and the agility to adapt to new challenges as they emerge. It requires the seamless integration of security practices into every facet of an organization's operations and culture.

Throughout this chapter, we will explore a range of strategies and best practices that embody the proactive cybersecurity approach. Each section will provide valuable insights and actionable advice, empowering individuals and organizations to not just survive but also thrive in the ongoing battle against cyber threats. From the art of threat anticipation to the science of attack surface minimization, and from the imperative of continuous learning to the necessity of fostering a pervasive culture of security awareness, these tools will equip you with the confidence needed to navigate the ever-shifting cyber threat landscape.

We will also examine the broader implications of adopting a proactive cybersecurity mindset. This includes its potential impact on collective defense efforts, the shaping of future legislation, and the evolution of regulatory frameworks. By understanding and effectively implementing proactive measures, organizations can protect their assets and contribute significantly to creating a safer and more resilient digital ecosystem for all.

The journey toward a genuinely proactive cybersecurity posture is one of commitment to continuous improvement and collaboration. It calls for leaders to elevate security to a top strategic priority, for employees at all levels to remain vigilant and well-informed, and for cybersecurity professionals to constantly push the boundaries of their expertise to stay ahead of evolving threats. This collaborative effort is crucial, and as we delve deeper into these crucial aspects, we aim to arm you with the knowledge, strategies, and mindset needed to navigate the ever-shifting cyber threat landscape with confidence and efficacy.

Join us as we embark on an exploration of the proactive cybersecurity approach. We'll unpack its fundamental principles, learn from real-world scenarios, and uncover practical steps that can be immediately implemented to enhance your organization's security posture. By embracing this forward-thinking mindset, we can collectively forge a path toward a more secure digital future where we are not merely reactive but also proactively shaping the cybersecurity landscape. Are you ready to take the first step toward a proactive cybersecurity approach?

The following topics are covered in this chapter:

- The need for a proactive approach

- Continuous learning and adaptation

- Fostering a proactive cybersecurity culture

- The broader implications of proactive cybersecurity

- The ongoing importance of cybersecurity

The need for a proactive approach

A proactive approach to cybersecurity is grounded in the principle of being one step ahead of potential threats. Instead of merely reacting to security incidents, organizations and individuals strive to anticipate and prepare for emerging threats before they manifest. Such an approach requires a blend of comprehensive threat intelligence, diligent risk assessment, and strategic planning:

- **Anticipating emerging threats**: Proactive cybersecurity begins with the anticipation of emerging threats. In the cyber realm, threats are not static; they continuously evolve. Cybercriminals and threat actors are persistent and innovative, perpetually seeking new ways to exploit vulnerabilities. Therefore, relying on past defense strategies is no longer sufficient.

  Anticipating threats involves staying well-informed about current threat landscapes, understanding the **tactics, techniques, and procedures** (**TTPs**) employed by threat actors, and closely monitoring emerging vulnerabilities. Cyber threat intelligence plays a crucial role in this process, providing valuable insights into evolving threats.

- **Minimizing attack surfaces**: A proactive approach to cybersecurity is also about minimizing attack surfaces. Attack surfaces refer to the points of entry and vulnerabilities within an organization's systems and networks. The more attack surfaces an organization has, the higher the likelihood of exploitation.

  Minimizing attack surfaces involves diligent configuration management and regular vulnerability assessments. System administrators need to ensure that systems are configured securely, and they must continuously monitor for vulnerabilities. Reducing the number of entry points for potential attackers mitigates the risk of exploitation.

- **Preparing for incidents**: Being proactive in cybersecurity means preparing for inevitable security incidents, such as data breaches or system compromises, by developing well-defined incident response procedures and policies. This preparation involves equipping an organization with the necessary forensic tools and services, including forensic workstations equipped with software such as EnCase or FTK. These tools provide a sense of control, as they are essential for analyzing compromised systems and preserving evidence. Additionally, having logs from critical systems readily available is crucial for investigations. Many organizations use centralized logging solutions such as **security information and event management** (**SIEM**) systems to consolidate and monitor logs from various sources, making it easier to detect patterns and conduct thorough investigations when incidents occur.

Moreover, engaging external consultants or vendors can be vital to a robust incident response strategy. These external experts can offer specialized skills and an objective perspective during complex incidents. For instance, a financial institution may partner with cybersecurity firms to perform forensic analysis during advanced attacks. At the same time, a healthcare provider might maintain an incident response retainer to address potential data breaches quickly. However, it's not just about the external support. Regular tabletop exercises, such as simulating ransomware attacks, are equally important. These exercises test response plans, ensure that team members are familiar with their roles, and identify procedural gaps, making an organization feel prepared and confident in its incident response strategy. By taking these proactive steps, organizations can minimize the impact of incidents, accelerate recovery, and protect their assets and reputation.

Continuous learning and adaptation

A proactive approach to cybersecurity is not static; it involves continuous learning and adaptation. Cyber threats and the methods employed by cybercriminals are constantly changing. To effectively defend against these evolving threats, individuals and organizations must stay informed, adapt to new challenges, and continually enhance their security practices:

- **Training and education**: Promoting continuous learning and training is essential for cybersecurity professionals. Technology and threat landscapes evolve rapidly, making it crucial to stay updated with the latest trends and risks. Regularly updated knowledge and skills are fundamental to staying abreast of evolving threats.

Professional certifications, such as **Certified Information Systems Security Professional (CISSP)** or **Certified Ethical Hacker (CEH)**, are important to demonstrate expertise and commitment to continuous learning.

- **Threat hunting**: Proactive threat hunting is a vital aspect of continuous learning and adaptation. Instead of waiting for security incidents to trigger alarms, organizations employ threat hunters to actively search for signs of malicious activity within their networks.

  Threat hunters rely on advanced knowledge and tools to identify subtle indicators of compromise. They meticulously analyze network traffic, logs, and endpoints to uncover threats that may go unnoticed by traditional security measures. Threat hunting is about being one step ahead, proactively identifying and neutralizing threats before they become major incidents.

- **Security awareness programs**: For organizations, promoting a proactive cybersecurity mindset among employees is paramount. Security awareness programs teach employees how to recognize and report potential security incidents and threats. They empower individuals to be the first line of defense.

  These programs provide employees with the knowledge and skills to identify phishing attempts, suspicious emails, and social engineering tactics. By fostering a culture of security awareness, organizations ensure that every member of a team proactively contributes to an organization's security posture.

To maintain a robust defense, continuous learning and adaptation are essential in navigating the ever-evolving cybersecurity landscape. Fostering a proactive cybersecurity culture empowers organizations to stay ahead of emerging threats by integrating security awareness into every layer of the business.

Fostering a proactive cybersecurity culture

Promoting a proactive cybersecurity mindset and continuous learning is not confined to individual efforts. It requires building a culture within organizations that embraces proactive security measures. The foundation for such a culture should be built from the top down, starting with leadership:

- **Leadership commitment**: A proactive cybersecurity culture begins with leadership, but it's about more than just having the right team, budget, or numerous vendors at their disposal. The success of an organization's cybersecurity efforts hinges on the mindset and attitude of its leaders toward security. Top executives, including the C-suite and the board of directors, must demonstrate a genuine commitment to security, prioritizing it as a core value rather than an afterthought. However, even with resources, leaders must have the right mindset to view security as an integral aspect of the organization's success. Otherwise, it can undermine the entire cybersecurity strategy.

  Investing in programs and awareness sessions designed explicitly for leadership is crucial to cultivating the right mindset. These initiatives can help leaders develop a holistic understanding of cybersecurity, emphasizing its importance beyond compliance and technology. Training leadership to view security as a strategic enabler, rather than a cost or hindrance, fosters a culture where cybersecurity is embedded into every aspect of an organization's operations. By shaping leadership attitudes through continuous education and awareness, organizations can ensure that their security posture is supported by leaders committed to protecting the business from emerging threats.

Leaders should invest in cybersecurity resources, allocate budgets, and actively engage in discussions about security strategy. Their commitment reinforces the importance of proactive security measures.

- **Integration of security**: A proactive cybersecurity culture integrates security into all aspects of an organization's operations. Security should not be seen as a standalone function but as an inherent component of everything the organization does. This entails considering security from the inception of new projects, products, or services.

  For example, security should be integrated into the **software development life cycle (SDLC)**. The concept of *security by design* ensures that security is considered at every stage of development, from design and coding to testing and deployment. Such integration is vital in identifying and addressing vulnerabilities early in the software design process.

The broader implications of proactive cybersecurity

A proactive cybersecurity mindset has broader implications, not just for individual organizations but also for the cybersecurity ecosystem as a whole. It contributes to a safer and more resilient digital environment in the following ways:

- **Collective defense**: A proactive approach to cybersecurity extends beyond individual organizations and fosters a sense of collective defense. Organizations, government agencies, and cybersecurity professionals collaborate to share threat intelligence, best practices, and mitigation strategies.

  Collective defense is especially critical when facing sophisticated and persistent cyber threats, such as nation-state-sponsored attacks. By sharing information and working together proactively, the cybersecurity community can effectively counteract these threats.

- **Legislation and regulation**: Proactive cybersecurity practices can influence legislation and regulation. Governments worldwide are enacting cybersecurity laws to safeguard critical infrastructure and sensitive data. A proactive cybersecurity culture encourages compliance with these laws and, in some cases, goes beyond compliance to exceed minimum requirements.

  By adhering to proactive security standards and practices, organizations can navigate the complex regulatory landscape with greater ease, avoid penalties, and enhance their reputation as responsible data custodians.

A proactive cybersecurity mindset is not merely a strategy; it is also a philosophy that underpins a resilient and adaptable approach to security. It is a commitment to staying one step ahead of evolving threats, minimizing vulnerabilities, and fostering a culture that values security at every level of an organization.

The significance of such a mindset extends beyond individual organizations. It influences collective defense efforts, shapes cybersecurity legislation and regulation, and ultimately contributes to a safer digital environment.

By encouraging a proactive cybersecurity mindset and investing in continuous learning and adaptation, organizations and individuals become not just defenders but also leaders in the ongoing battle against cyber threats. They epitomize the principle that the best way to predict the future of cybersecurity is to create it.

The ongoing importance of cybersecurity

The significance of cybersecurity has never been more apparent. This section underscores the role of cybersecurity in an increasingly digital world, highlighting the reasons why it remains a paramount concern.

The digital transformation

It's crucial to understand the broader context before diving into the specific aspects of digital transformation and its impact on cybersecurity. As we transition into an increasingly connected world, the following vital areas highlight how digital transformation influences cybersecurity and why it demands a proactive approach.

- **Ubiquitous connectivity**: The world is more connected than ever before. With the proliferation of the **Internet of Things** (**IoT**), mobile devices, and cloud computing, almost every aspect of our lives is linked to the digital realm. This interconnectedness enhances convenience, efficiency, and accessibility to information. However, it also broadens the attack surface for cyber threats. The IoT, for instance, has brought connectivity to everyday objects, from thermostats and refrigerators to industrial machinery. While this connectivity offers unparalleled convenience and automation, it also presents new vulnerabilities. Poorly secured IoT devices can be exploited by malicious actors to gain access to networks, data, or even control over physical systems. As these devices become more integral to daily life, the need for robust security measures becomes increasingly critical.

  Additionally, the widespread use of mobile devices has led to a blurring of the lines between personal and professional computing. Employees often use their own devices for work-related tasks, creating a phenomenon known as **Bring Your Own Device** (**BYOD**). While BYOD can boost productivity, it also introduces security challenges. Organizations must grapple with securing a diverse array of devices and ensuring that sensitive corporate data is protected, no matter where it is accessed.

  Cloud computing, which offers scalable and cost-effective storage and computing resources, has also transformed the way businesses operate. However, the transfer of data and applications to the cloud necessitates robust security measures to safeguard sensitive information. Cloud security involves encryption, access control, and monitoring, among other strategies.

- **Data as a currency**: Data has become a valuable currency in the digital economy. Personal, financial, and business data are all targets for cybercriminals. Protecting this data is crucial not only for individual privacy but also for economic and national security. The collection, storage, and analysis of data underpin many aspects of modern life. Individuals entrust their personal information to various online platforms, from social media to e-commerce sites. This data includes personal details, preferences, and financial information. Protecting this data is vital, as data breaches can result in identity theft, financial losses, and personal privacy violations.

In the business world, data is not just an asset but also often a competitive advantage. Companies collect and analyze data to gain insights into customer behavior, streamline operations, and make informed decisions. This data includes customer records, proprietary research, and financial information. For organizations, a data breach can result in significant financial losses, damage to reputation, and legal liabilities.

Moreover, data plays a pivotal role. Governments collect vast amounts of data for intelligence and security purposes. Protecting this data is not only essential for safeguarding a nation's interests but also for preserving international relations and alliances. Breaches of national security databases can have far-reaching consequences, including diplomatic tensions and potential threats to a country's citizens.

Cyber threat landscape

Threat actors have become more sophisticated and diverse in the evolving cyber landscape. Understanding who these actors are and how they operate is essential to defending against their attacks. Next, we will explore the different types of threat actors and the emerging vectors they exploit:

- **Sophisticated threat actors**: The world of cybercrime has evolved to include highly sophisticated threat actors. These include nation-states, organized crime groups, hacktivists, and even terrorist organizations. What sets these actors apart is their substantial resources, advanced technical capabilities, and the ability to launch complex, large-scale attacks. Nation-states, for instance, have recognized the potential of cyber warfare and have invested heavily in building cyber capabilities. These well-funded entities possess the resources to develop highly advanced malware, conduct espionage, and launch cyberattacks on critical infrastructure. They operate with strategic objectives and can disrupt critical services and influence international affairs.

  Organized crime groups are driven by financial motivations. They engage in activities such as ransomware attacks, data theft, and fraud. Their operations are often well-organized, with a focus on maximizing profit. They employ a range of tactics, from phishing campaigns to exploit kits.

  Conversely, hacktivists use cyberattacks to advance their political or social agendas. They may target government websites, corporations, or other entities they perceive as adversaries. Their attacks can range from website defacement to data breaches aimed at exposing sensitive information.

- **Emerging threat vectors**: New attack vectors continue to emerge, challenging traditional security measures and necessitating advanced defense strategies:

- **Supply chain attacks**: Supply chain attacks involve targeting a software or hardware supply chain to infiltrate organizations. Attackers compromise a trusted vendor's products, which are then distributed to target organizations. Notable examples include the SolarWinds hack and the NotPetya malware, both of which leveraged supply chain weaknesses.

- **Artificial Intelligence (AI)-driven attacks**: The use of AI in cyberattacks is on the rise. AI can be employed to automate tasks, such as credential stuffing attacks, or to enhance the precision of phishing campaigns. It can also be used for evasive maneuvers in response to security measures.

Implications of cyber threats

Cyber threats have far-reaching implications that extend beyond individual organizations. These implications can affect the economy, national security, and society. The following sections detail the significant impacts of cyber threats and why they require a robust and proactive defense:

- **Economic impact**: Cyberattacks have profound economic consequences. They can lead to financial losses, damage to an organization's reputation, and legal liabilities. The broader economy can also be affected when major incidents disrupt critical infrastructure. The direct financial impact of cyberattacks can be substantial. Costs include expenses related to incident response, legal fees, regulatory fines, and the recovery of compromised systems. In the case of ransomware attacks, organizations may be forced to pay substantial sums to regain access to their data.

  Successful cyberattacks can severely damage an organization's reputation. The erosion of trust among customers, partners, and stakeholders can result in lost business and long-term financial consequences.

  Organizations may face regulatory penalties and legal actions following data breaches, especially if they fail to comply with data protection laws. Legal proceedings and settlements can incur significant costs.

- **National security**: Cybersecurity is a national security concern. Nation-states have recognized the potential of cyber warfare and have invested heavily in building cyber capabilities. The ability to protect national interests in the digital domain is closely tied to cybersecurity. Critical infrastructure, such as power grids, water supply systems, and transportation networks, relies on digital systems for operation and management. A successful cyberattack on critical infrastructure can have catastrophic consequences, including power outages, environmental disasters, and disruptions to essential services. Cyber warfare capabilities are integrated into the strategies of many nations. These capabilities include offensive cyber operations aimed at disrupting the communications and logistics of adversaries. Defending against such attacks is crucial for national security.

The role of cybersecurity

Cybersecurity plays a pivotal role in the modern digital landscape, serving multifaceted purposes, ranging from protecting critical infrastructure to preserving individual privacy. Let's delve into these key roles in more detail:

- **Protecting critical infrastructure**: Critical infrastructure comprises systems and assets that are vital to the functioning of a society and its economy. This includes energy grids, water supply systems, and transportation networks. These essential services often rely on digital control systems to ensure their efficient operation. However, the integration of digital technologies into critical infrastructure also exposes them to the ever-present threat of cyberattacks. The consequences of a successful cyberattack on critical infrastructure can be catastrophic. An attacker gaining control over the energy grid could disrupt the power supply to entire regions, leading to power outages, economic losses, and potential threats to public safety. Water supply systems are not immune to cyber threats either; unauthorized access could contaminate water sources, posing health risks to the population.

  Transportation networks, including air traffic control and railway systems, also face cyber risks. Disruptions in these systems can result in accidents, causing loss of life and extensive damage. As such, cybersecurity is not a mere option but a critical imperative to protect the very backbone of modern society.

- **Individual privacy**: The proliferation of digital services and platforms has led to the generation and exchange of vast amounts of personal information. Protecting individual privacy has become a matter of paramount importance. Cybersecurity measures are indispensable in this context, ensuring that sensitive data remains confidential and secure. One of the fundamental elements of safeguarding individual privacy in the digital age is encryption. Encryption technologies, such as SSL/TLS for secure web connections, protect data during transmission. End-to-end encryption in messaging applications ensures that only the intended recipient can decipher the messages, preserving the confidentiality of personal communication.

Data protection regulations, such as the **European Union's General Data Protection Regulation (GDPR)** and the **California Consumer Privacy Act (CCPA)**, have been enacted to enforce individual privacy rights and place stringent requirements on organizations to protect personal data. Compliance with these regulations has become a legal obligation, emphasizing the significance of cybersecurity in the digital era.

Future challenges and solutions

As the cybersecurity landscape evolves, it's essential to anticipate and prepare for future challenges while seeking innovative solutions. Here are some key considerations:

- **AI and Machine Learning (ML)**: AI and ML have become integral components of modern cybersecurity. These technologies are not only employed by threat actors but also offer significant potential to enhance cybersecurity. AI-driven security tools can identify and respond to threats in real time, providing a proactive defense. Behaviour-based anomaly detection, as mentioned earlier, is one of the prime applications of AI and ML in cybersecurity. These systems scrutinize network and user behaviors, identifying deviations that may indicate cyberattacks. They excel at recognizing novel threats, such as zero-day exploits, by learning normal behavior and detecting outliers.

Machine learning models are also instrumental in advanced malware detection. By analyzing code behavior, file attributes, and network activities, they can identify previously unknown malware. This proactive approach significantly bolsters an organization's ability to defend against evolving threats.

- **Quantum computing**: The development of quantum computing represents a potential challenge to current encryption methods. Quantum computers, when fully realized, could potentially break existing encryption standards. This paradigm shift in computing necessitates the development of quantum-resistant encryption techniques to counteract this emerging threat. Researchers and cybersecurity experts are actively engaged in this field, working to devise encryption methods that can withstand the computational power of quantum computers. Organizations must stay proactive in this area, closely monitoring developments in quantum computing and adapting their security measures accordingly to ensure the long-term confidentiality and integrity of their data.

- **International collaboration**: Cyber threats transcend national borders, making international cooperation imperative to address them effectively. Countries, organizations, and cybersecurity professionals worldwide work together to share threat intelligence, develop international norms, and promote cybersecurity best practices. The exchange of threat intelligence enables organizations to stay informed about emerging threats and vulnerabilities. By sharing information about cyber incidents and threat indicators, global entities can collectively enhance their cybersecurity posture and respond to cyber threats more effectively.

Developing international norms and best practices sets the foundation for a unified approach to cybersecurity. Agreements and guidelines help create a common framework to respond to cyber incidents and promote responsible behavior in cyberspace.

The role of cybersecurity extends beyond protecting data; it includes safeguarding critical infrastructure, preserving individual privacy, and ensuring national security. As technology continues to advance, new challenges will arise, demanding proactive solutions and a commitment to staying ahead of emerging threats.

For individuals, organizations, and nations, recognizing the critical importance of cybersecurity and acting proactively to address the evolving landscape is not just a choice; it is also an imperative. The significance of cybersecurity will only grow as our world becomes more interconnected, making it a fundamental element of the digital age.

Time to take action

As we conclude this exploration of the cyber kill chain and the multifaceted world of cybersecurity, it's essential to underscore the critical importance of taking action. Cyber threats are not abstract concepts; they are real, dynamic, and potentially devastating. Whether you are an individual, a business leader, or a cybersecurity professional, you have a role to play in safeguarding the digital landscape.

In an interconnected world, cybersecurity is a shared responsibility. Cyber threats have a broad impact, and the consequences of inaction can ripple through society. Recognizing this shared responsibility is the first step toward proactive defense. Let's look into what responsibilities an individual, a business, and a cybersecurity professional can adopt to improve cybersecurity.

For individuals

As individuals, we play a crucial role in cybersecurity, albeit on a smaller scale than organizations or professionals. Our actions can protect personal data and prevent cyber threats from spreading:

- **Be informed**: Stay informed about current cyber threats and best practices for personal cybersecurity. Follow reputable sources for cybersecurity news, and consider enrolling in online courses or workshops to enhance your knowledge.

- **Practice good cyber hygiene**: Secure your devices and online accounts with strong, unique passwords, enable two-factor authentication where possible, and be cautious about clicking on links or downloading attachments from unknown sources.

- **Stay vigilant**: Be on the lookout for common cyber threats such as phishing emails and fraudulent websites. Report suspicious activity to your organization's IT department or the appropriate authorities.

For businesses and organizations

Organizations have a broader responsibility in the cybersecurity landscape. Their actions can protect not only their assets but also their customers, partners, and the wider economy:

- **Invest in cybersecurity**: Allocate resources to build a robust cybersecurity program. This includes investing in security technologies, training your employees, and regularly assessing and improving your security posture.

- **Develop an incident response plan**: Prepare for the inevitability of security incidents by creating an incident response plan. This plan should outline procedures to identify, contain, and mitigate breaches.

- **Implement a zero trust model**: Adopt the zero trust security model, which assumes that threats exist both inside and outside a network. It requires strict identity verification for anyone trying to access resources on the network.

For cybersecurity professionals

Cybersecurity professionals are at the forefront of defending against cyber threats. Their expertise and proactive measures can significantly impact the security posture of the organizations they serve:

- **Stay current**: Continuously update your skills and knowledge. Cyber threats evolve rapidly, and staying current is essential for effective defense.

- **Engage in threat hunting**: Proactive threat hunting is a valuable practice for cybersecurity professionals. It involves actively searching for signs of malicious activity within your organization's network.

- **Share intelligence**: Collaborate with the broader cybersecurity community. Share threat intelligence, best practices, and mitigation strategies to collectively defend against cyber threats.

It's time to take action. The proactive approach to cybersecurity involves anticipating threats, minimizing vulnerabilities, and fostering a culture of security awareness.

Cybersecurity is not a choice; it's a necessity. It's the foundation upon which our digital lives, our businesses, and our society rest. It's the promise of a safer and more secure future.

Let us be the architects of this future. Let us embrace the proactive mindset, adapt to evolving threats, and collaborate with others to build a resilient defense against the cyber kill chain. The time for action is now.

Together, we can strengthen our digital world against cyber threats, ensuring a safer, more secure tomorrow for all.

Summary

This chapter on a proactive cybersecurity approach underscored the urgent need for a proactive stance in the face of the growing complexity and frequency of cyber threats. It advocated for a forward-thinking mindset that goes beyond reactive measures, urging organizations to anticipate and prepare for potential threats before they materialize. This approach, which involves a blend of comprehensive threat intelligence, diligent risk assessment, and strategic planning, is crucial to staying ahead of cybercriminals.

A crucial aspect of the proactive approach is the strategic reduction of attack surfaces. Organizations need to be well-informed about the current threat landscape, understand the tactics and procedures used by threat actors, and closely monitor for new vulnerabilities. By leveraging cyber threat intelligence, organizations can gain valuable insights into evolving threats, allowing them to prepare and implement measures to mitigate risks effectively. However, it's equally important to reduce attack surfaces through secure configurations and regular vulnerability assessments, as this is a key strategy in minimizing entry points for attackers.

Another critical component is the preparation for inevitable security incidents. This chapter stressed the importance of having well-defined incident response procedures and policies. These procedures outline the steps to detect, contain, eradicate, and recover from security incidents, ensuring a swift and coordinated response that minimizes the impact on the organization. This preparation provides a sense of security, and developing an incident response playbook is recommended to enhance readiness and effectiveness in handling breaches.

Finally, the chapter highlighted the significance of continuous learning and fostering a proactive cybersecurity culture. Cybersecurity professionals must undergo ongoing training and education to stay abreast of the latest trends and threats. This continuous learning keeps you engaged and includes obtaining professional certifications and participating in proactive threat-hunting activities. Furthermore, building a security culture within organizations requires leadership commitment and integrating security practices into all operations. Organizations can better defend against evolving cyber threats and contribute to a safer digital environment by promoting a proactive mindset and continuous adaptation.

Thank you for completing *Cyber Kill Chain – Tactics and Strategies*. We appreciate your dedication and effort in exploring the intricate world of cybersecurity through this book. The journey through the cyber kill chain has provided a comprehensive understanding of the various stages of cyberattacks, from reconnaissance to taking actions on objectives, and the critical importance of legal and ethical considerations in defending against these threats.

Throughout this book, we delved into the evolution of the cyber kill chain model, the complexities of **advanced persistent threats** (**APTs**), and the integration of emerging technologies such as **operational technology** (**OT**) and the IoT. We also examined the critical role of international cooperation, legal enforcement mechanisms, and ethical frameworks in building a robust cybersecurity posture.

As you continue your journey in the field of cybersecurity, we encourage you to stay informed about the latest developments and advancements. Cybersecurity is a dynamic and ever-evolving field, and continuous learning is essential to stay ahead of emerging threats and challenges. We recommend exploring additional resources, such as industry reports, academic journals, and online courses, to further deepen your knowledge and expertise.

Further reading

Following are some bonus reading materials for you to check out:

- *An overview of the Cyber Kill Chain framework and its seven phases for understanding and defending against cyberattacks*: `https://www.crowe.com/cybersecurity-watch/building-defenses-with-the-cyber-kill-chain-framework`

- *A comprehensive guide to cyber kill chains, including their origins, components, and application in modern cybersecurity strategies*: `https://www.splunk.com/en_us/blog/learn/cyber-kill-chains.html`

- *Discusses the top five challenges organizations face when implementing the Cyber Kill Chain framework and how to overcome them*: `https://kravensecurity.com/top-5-cyber-kill-chain-challenges/`

- *An in-depth explanation of the Cyber Kill Chain, its importance in cybersecurity, and how it can be applied to enhance threat intelligence*: `https://www.recordedfuture.com/threat-intelligence-101/threat-analysis/cyber-kill-chain`

- *A white paper by Lockheed Martin detailing seven ways to apply the Cyber Kill Chain with a Threat Intelligence Platform*: `https://www.lockheedmartin.com/content/dam/ lockheed-martin/rms/documents/cyber/Seven_Ways_to_Apply_the_ Cyber_Kill_Chain_with_a_Threat_Intelligence_Platform.pdf`

- *A guide to understanding and applying the Lockheed Martin Kill Chain, including real-world case studies and tips for strengthening cybersecurity*: `https://www.mycloudsec. com/demystifying-the-lockheed-martin-kill-chain-a-guide-to- strengthening-cybersecurity`

Subscribe to _secpro – the newsletter read by 65,000+ cybersecurity professionals

Want to keep up with the latest cybersecurity threats, defenses, tools, and strategies?

Scan the QR code to subscribe to **_secpro**—the weekly newsletter trusted by 65,000+ cybersecurity professionals who stay informed and ahead of evolving risks.

`https://secpro.substack.com`

Get This Book's PDF Version and Exclusive Extras

UNLOCK NOW

Scan the QR code (or go to `packtpub.com/unlock`). Search for this book by name, confirm the edition, and then follow the steps on the page.

Note: Keep your invoice handly. Purchase made directly from Packt don't require one.

13

Unlock Your Exclusive Benefits

Your copy of this book includes the following exclusive benefits:

- ⌂ Next-gen Packt Reader
- 📄 DRM-free PDF/ePub downloads

Follow the guide below to unlock them. The process takes only a few minutes and needs to be completed once.

Unlock this Book's Free Benefits in 3 Easy Steps

Step 1

Keep your purchase invoice ready for *Step 3*. If you have a physical copy, scan it using your phone and save it as a PDF, JPG, or PNG.

For more help on finding your invoice, visit `https://www.packtpub.com/unlock-benefits/help`.

> **Note**
> If you bought this book directly from Packt, no invoice is required. After *Step 2*, you can access your exclusive content right away.

Step 2

Scan the QR code or go to `packtpub.com/unlock`.

On the page that opens (similar to *Figure 13.1* on desktop), search for this book by name and select the correct edition.

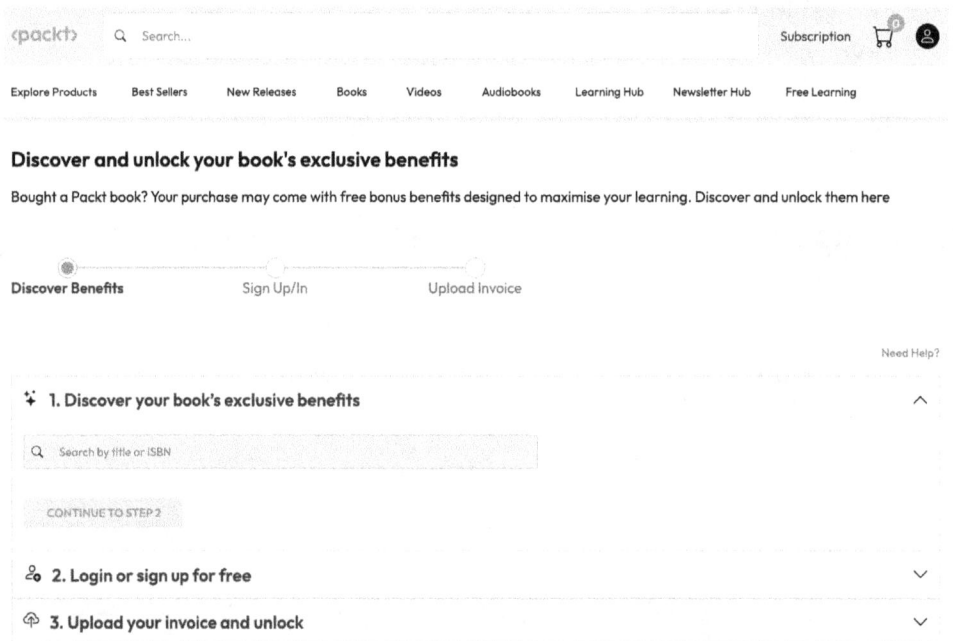

Figure 13.1: Packt unlock landing page on desktop

Step 3

After selecting your book, sign in to your Packt account or create one for free. Then upload your invoice (PDF, PNG, or JPG, up to 10 MB). Follow the on-screen instructions to finish the process.

Need help?

If you get stuck and need help, visit
`https://www.packtpub.com/unlock-benefits/help`
for a detailed FAQ on how to find your invoices and more. This QR
code will take you to the help page.

Note

If you are still facing issues, reach out to `customercare@packt.com`.

Index

‹packt›

packt.com

Subscribe to our online digital library for full access to over 7,000 books and videos, as well as industry leading tools to help you plan your personal development and advance your career. For more information, please visit our website.

Why subscribe?

- Spend less time learning and more time coding with practical eBooks and Videos from over 4,000 industry professionals
- Improve your learning with Skill Plans built especially for you
- Get a free eBook or video every month
- Fully searchable for easy access to vital information
- Copy and paste, print, and bookmark content

At www.packt.com, you can also read a collection of free technical articles, sign up for a range of free newsletters, and receive exclusive discounts and offers on Packt books and eBooks.

Other Books You May Enjoy

If you enjoyed this book, you may be interested in these other books by Packt:

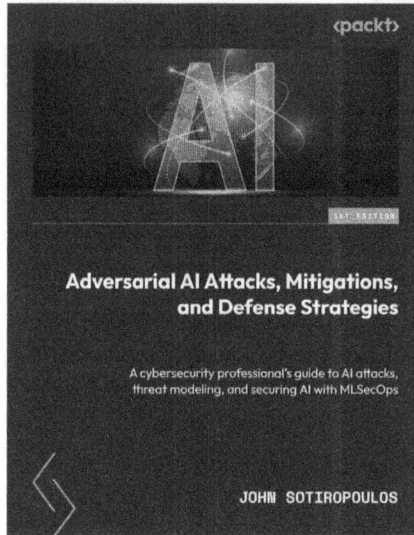

Adversarial AI Attacks, Mitigations, and Defense Strategies

John Sotiropoulos

ISBN: 978-1-83508-798-5

- Set up a playground to explore how adversarial attacks work
- Discover how AI models can be poisoned and what you can do to prevent this
- Learn about the use of trojan horses to tamper with and reprogram models
- Understand supply chain risks
- Examine how your models or data can be stolen in privacy attacks
- See how GANs are weaponized for Deepfake creation and cyberattacks
- Explore emerging LLM-specific attacks, such as prompt injection

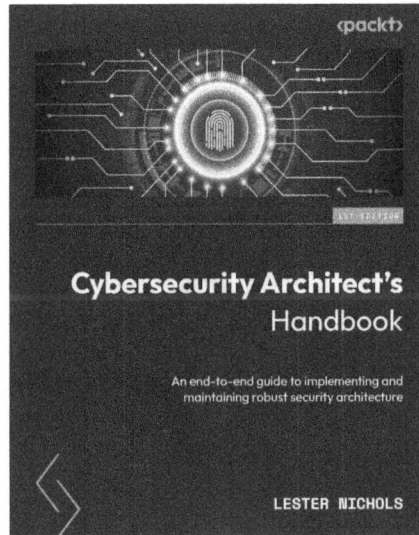

Cybersecurity Architect's Handbook

Lester Nichols

ISBN: 978-1-80323-584-4

- Get to grips with the foundational concepts and basics of cybersecurity
- Understand cybersecurity architecture principles through scenario-based examples
- Navigate the certification landscape and understand key considerations for getting certified
- Implement zero-trust authentication with practical examples and best practices
- Find out how to choose commercial and open source tools
- Address architecture challenges, focusing on mitigating threats and organizational governance

> **Note**
>
> Looking for more cybersecurity books? Browse our full catalog at: `https://packt.link/cybersecurity`.

Packt is searching for authors like you

If you're interested in becoming an author for Packt, please visit `authors.packtpub.com` and apply today. We have worked with thousands of developers and tech professionals, just like you, to help them share their insight with the global tech community. You can make a general application, apply for a specific hot topic that we are recruiting an author for, or submit your own idea.

Share your thoughts

Now you've finished *Cyber Security Kill Chain - Tactics and Strategies*, we'd love to hear your thoughts! Scan the QR code below to go straight to the Amazon review page for this book and share your feedback or leave a review on the site that you purchased it from.

`https://packt.link/r/1835466095`

Your review is important to us and the tech community and will help us make sure we're delivering excellent quality content.

‹packt› _secpro

Stay relevant in a rapidly changing cybersecurity world – join 65,000+ SecPro subscribers

_secpro is the trusted weekly newsletter for cybersecurity professionals who want to stay informed about real-world threats, cutting-edge research, and actionable defensive strategies.

Each issue delivers high-signal, expert insights on topics like:

- Threat intelligence and emerging attack vectors
- Red and blue team tactics
- Zero Trust, MITRE ATT&CK, and adversary simulations
- Security automation, incident response, and more!

Whether you're a penetration tester, SOC analyst, security engineer, or CISO, **_secpro** keeps you ahead of the latest developments — no fluff, just real answers that matter.

Scan the QR code to subscribe for free and get expert cybersecurity insights straight to your inbox:

https://secpro.substack.com